ANALYTICAL CHEMISTRY SYMPOSIA SERIES — volume 15

computer applications in chemistry

Proceedings of the 6th International Conference on Computers in Chemical Research and Education (ICCCRE), held in Washington, DC, July 11–16, 1982

ANALYTICAL CHEMISTRY SYMPOSIA SERIES — volume 15

computer applications in chemistry

Proceedings of the 6th International Conference on Computers in Chemical Research and Education (ICCCRE), held in Washington, DC, July 11—16, 1982

edited by

Stephen R. Heller and Rudolph Potenzone, Jr.

U.S. Environmental Protection Agency, Washington, DC, U.S.A.

ELSEVIER — Amsterdam — Oxford — New York 1983

ELSEVIER SCIENCE PUBLISHERS B.V.
Molenwerf 1
P.O. Box 211, 1000 AE Amsterdam, The Netherlands

Distributors for the United States and Canada:

ELSEVIER SCIENCE PUBLISHING COMPANY INC.
52, Vanderbilt Avenue
New York, NY 10017

Library of Congress Cataloging in Publication Data

International Conference on Computers in Chemical
    Research, Education, and Technology (6th : 1982 :
    Washington, D.C.)
    Computer applications in chemistry.

    (Analytical chemistry symposia series ; v. 15)
    Includes indexes.
    1. Chemistry--Data processing--Congresses. I. Heller,
Steven R., 1943-   . II. Potenzone, Rudolf, 1952-   .
III. Title. IV. Series.
QD39.3.E46I56 1982   542'.8   83-9009
ISBN 0-444-42210-2

ISBN 0-444-42210-2 (Vol. 15)
ISBN 0-444-41786-9 (Series)

Printed in The Netherlands

CONTENTS

ANALYTICAL CHEMISTRY SYMPOSIA SERIES

Volume 15   Computer Applications in Chemistry. Proceedings of the 6th International Conference on Computers in Chemical Research and Education (ICCCRE), held in Washington, DC, July 11—16, 1982
edited by S.R. Heller and R. Potenzone, Jr.

Dedicated to Morris B. Yaguda

INTRODUCTION

This book represents an almost complete summary of the plenary lectures and poster session presentations given at the 6th International Conference on Computers in Chemical Research and Education (ICCCRE), held in Washington DC from 11-16 July 1982.

The format for the 6th ICCCRE differed from the previous meetings, in that there were 18 main plenary lectures and some 70 poster talks. The plenary lectures were held in the mornings and evenings to give the participants time for informal discussions. Furthermore, lengthy coffee breaks added to the informal interactions at the meeting.

The growing use of computers in all areas of chemistry is becoming more evident every day. It is now critical that chemists become "computer literate" and understand what this powerful tool can do to assist in many chemical problems.

The members of the Local and International Organizing Committees provided much valuable advice and assistance in making the conference a scientific success. The International Committee members are Professor J. T. Clerc, Academican V. A. Koptyug, Professor P. Lykos, Professor S. Sasaki and Professor J. Zupan. The members of the Local Organzing Committee are Professor R. Dessy, Professor C. F. Hammer, Dr. G. W. A. Milne and Professor C. L. Wilkins, in addition to the editors of these proceedings.

The Organizing Committee of the 6th ICCCRE would like to express its deep appreciation to a number of organizations and individuals who assisted in making the meeting so highly sucessful. The sponsorship of Georgetown University, the US Environmental Protection Agency, and the International Union of Pure and Applied Chemistry (IUPAC) is particularly appreciated.

The following organizations provided financial assisatnce to allow for many participants in the USA and elsewhere to attend the conference. On behalf of the Organizing Committee and the participants who received some financial assistance we would like to thank the following:

EPA, Management Information and Data Systems Division
NCI, Information Technology Branch
Lederle Labs
Eastman Kodak Company
G. D. Searle
Merck & Co.
Finnigan MAT
Nicolet Analytical Instruments
Questel/DARC
CIS Inc.
Abbott Labs
Fraser-Williams Ltd.
SOHIO
Infometrix
Lockheed Information Systems
Molecular Design Ltd.
Schering Corporation
JEOL (USA) Analytical Instruments
ISI

We also wish to thank the Tektronix Corporation for providing
the 6th ICCCRE with a graphics terminal. The most excellent
coffee was provided by the Brazilian Coffee Institute. The soft
drinks, which helped make the extreme heat and high humidity of
Washington DC more bearable was kindly donated by The Pepsi-Cola
Bottler of Washington DC, Inc.

Particular appreciation must be given to the conference secretary, Mrs. Anne Rusiewicz, who also put this manuscript together
into one standard computer readable file from which this book
was then printed, giving quality and consistency to the chapters
and abstracts of the poster talks.

Lastly, our special thanks to Professor Charles F. Hammer of
Georgetown University who made sure everything ran smoothly.

                              Stephen R. Heller
                              Rudolph Potenzone, Jr.

                              Washington, DC
                              March 1983

S.R. Heller and R. Potenzone, Jr. (Editors), *Computer Applications in Chemistry*
© 1983 Elsevier Science Publishers B.V., Amsterdam — Printed in The Netherlands

THE NEGLECTED INGREDIENT IN CHEMICAL COMPUTER SYSTEMS

E. HYDE

Managing Director: Fraser Williams, (Scientific Systems) Ltd., Poynton, Cheshire, ENGLAND

INTRODUCTION

Most scientists are now well served by the computing facilities which are available to them in the laboratory. In fact at this moment we are finding that computer salesmen are like the automobile trade and are following fashion trends as well as technical advances. It would appear therefore to be a good time to examine an aspect of computing not associated with hardware or software. The compilation and indexing of chemical substances and their properties is an area neglected by practically all scientists.

It is important to examine the two distinct areas of computing in chemistry. The first area is mainly concerned with the volume of external publications and is the province of the online services and abstracting organisations. The resultant information files are the products of abstractors and indexers and the operating computer systems are mainly designed by the combined efforts of data processor personnel and information scientists. The objectives are to provide alerting services to users and to provide retrieval facilities for use with large archival files. Techniques applied throughout are designed to cope with the problem of volumes of information and to provide services to the largest possible range of users. The result is somewhat of a hit and miss affair. Structure searching has been introduced recently to overcome the obvious disadvantage of text based methods of searching chemical indexes. But this is a cosmetic action and does little to overcome the inadequate methods of indexing which is a fundamental fault common to many existing services. Furthermore structure methods are a mystery to the very large number of users of chemical information systems, who are non-chemists. Thus users of chemical

information   retrieval   services   are   inadequately   served   by
present systems.

The second area of computing of value to the chemist is that
of information analysis. Many laboratory computing systems are
used for this purpose.   Usually the data files are prepared
locally, often from information generated in-house and sometimes
by direct data acquisition systems.   Many studies, such as
structure   property   analysis   are   limited   to optimisation
techniques because the data available is limited to a particular
chemical series or a specific testing method. More fundamental
property correlations requiring the analysis of thousands of
chemicals and their property data, which may result in the
detection of otherwise unknown phenomena, are rarely carried out
due to the lack of access to validated data.

Laying the Blame

Before we examine the problems of abstracting, indexing and
storage of information, and become complacent that it is
obviously the fault of the information profession, let it be
stated quite clearly that the fault lies with the attitude of
the chemical profession. The chemical profession has a hold over
many other scientific professions especially in the medical and
biological areas.   For long enough chemists have enjoyed a
special place in the scheme of things. Their ability to create a
sign language and to converse freely in unpronounceable words
has enabled them to maintain considerable mystique over their
actions. Keeping information within a closed circuit can often
be of considerable value, and there is no doubt about the high
standard of communication both verbally and written, which
occurs within the chemical fraternity. Unfortunately most of
this is tied to the seemingly endless pursuit and fascination of
chemistry itself. This would be an admirable tendency if
chemistry did not have an enormous impact on both man and his
environment. We only have to recollect the tragedy of
Thalidomide and the present controversy over the 245T herbicides
to realise that the chemical profession must not only
communicate within its own ranks but has a duty to see that the
sum total of knowledge is recorded in a manner which makes that
knowledge available for dissemination and analysis by both
chemists and non chemists.

Problems with Chemical Data Bases

Chemists have created a high standard of professionalism in the publication of technical papers, review articles and books. It is not the quality of writing or the setting of standards by highly respected referees where the problem lies, but in the subsequent abstracting and indexing. The objective of the chemical profession to present new facts as rapidly as possible is often in conflict with the real problem of recovering those facts in the sheer size of the resulting compilation of volumes of abstracts. Presentation and publication of new information is an important aspect of the advancement of chemistry, and its early disclosure can often be justified. However, this part only deals with the alerting service and not the retrieval of these important facts when they are ultimately recorded in context with other important data. Information gained by the work of a chemist ages at a far lower rate than that of many other allied subjects. The retrieval of chemical information correctly classified within the framework of existing knowledge must also be a major objective of the profession, and this is not so at the present time. Practising chemists pay little attention to the indexing and methods of retrieval of information even when that information is for use by chemists. Chemists pay no attention to the problem of retrieval of information for use by non-chemists.

A number of facts can now be examined. The major retrieval tool of the chemical profession is Chemical Abstracts. For years their problem has been to cope with the phenomenal output of chemists, both in fundamental advances in chemistry and also with associated property information. The computer was brought in to provide help with the problems of processing and storing data and with the mechanisation of publications. Chemical Abstract Services adapted to this new technology early and exceedingly well. It enabled them to keep abreast of the publication needs especially in the production of indexes. However, the methods adopted were not appropriate to the problems of indexing and classification and in this they reflected the lack of interest of the chemical profession. The result is monolithic files with inadequate coordination of relevant information and no indication of the authenticity of data contained in the system.

If there is one way of summing up the present problems with
chemical data bases it is that they are indexer convenient
rather than user oriented. These problems centre around the use
of systematic nomenclature, the CA registry number and the
failure to subdivide files.

A most important error is the continued fascination with
systematic nomenclature when it is recognised by most practising
chemists that it is little understood and of even less practical
value. A certain amount of effort is devoted to the introduction
of classification by the use of inverted names, but this
technique can mislead both chemist and non chemist. And did we
really need the 9th Collective Index Name?

But perhaps the missed opportunities are the more serious.
The computer is designed to associate facts and has great
classification potentials. However, present chemical information
services have not attempted to exploit these facilities in the
way that services have recorded chemical compounds. Every
compound is unique, but its uses and properties have
relationships which are necessary to establish if we are to
fully utilise the recorded data. Before any compound is encoded
into a machine readable form it is important that it is first
examined against a set of indexing conventions. Yet the CA
registry number system records only the unique status of each
identified compound. This practice leads to the unnecessary
separation of essential data.   A typical example of this occurs
in the publicity sheet for the new edition of the Dictionary of
Organic Compounds shortly to be published.

Chlorophenitamine	132-22-9
(S) - form	25523-97-1
Maleate	119-92-8
(+/-) - form	42882-96-2

This shows clearly how the separation of compounds occur by the
lack of application of indexing rules. A simple logical form of
computer representation could have readily created an indexing
procedure which would have clearly related these molecules,
without any need to resort to structure searching techniques.
The compilers of the Dictionary of Organic Compounds concluded
that in a number of cases it was necessary to limit the number

of CA registry entries.

It is not sufficient to allow the entering or reentering of a compound into a retrieval file to be based solely on the occurrence of that compound in a paper.   Information on some compounds is of considerably more value to users than the knowledge that yet another compound exists. Volume in itself is not necessarily a virtue. There are 5 1/2 million compounds, but the whole chemical industry is founded on less than 100,000 compounds. Many users of chemical information systems require pertinent facts on a much smaller collection than 5 1/2 million, and it should be possible to enter a search in a preferred search level before having to enter into the 5 1/2 million compound file.

Thus a number of problems require further examination if there is to be established a chemical information data base which better fulfills the need for comprehensive retrieval, for information analysis and suitable for use by non-chemists.

Improving Chemical Information

The first step in improving chemical data bases is to establish indexing conventions. Very little attention is paid to the classification of compounds in a manner which will give total recall of all relevant data. This problem not only affects parent and salts, but forms, spatial configurations, water of crystallisation and internal salts. Most in house systems and some external data base producers allow the recording of molecules exactly as they are presented to the encoder. Encoding is mistaken for indexing. There appears to be little recognition that indexing must precede encoding if satisfactory retrieval is to be attained. A logical way of organising a chemical record can be devised which after applying the indexing rules will enable encoding to take place in a manner which will achieve the required association between compounds.

One such method is to divide the compound record into five parts as follows.

Logical Compound Record composed of
    1. Two Dimensional Network   -Parent
    2. Modifications to the network   -Na salt
    3. Additions to the network   -HCl

4. Spatial Information   -Third dimension
5. Proportion of Constituent present -Purity

The logical ordering of a chemical record can then be used
to produce a hierarchical number which will produce the
necessary association without the loss of a unique identifier
for each compound.   Relationships can readily be established
between parents and salts, non stereo and stereo forms, and pure
compounds   and   their   occurrence   in   mixtures.   There are
considerable benefits to both chemists and non-chemists on
basing a registry number on such a procedure. A unique number
entered as a query would immediately alert the user to the other
compounds held with the same parent number.   Advantages also
occur in substructure searching.   The number of records to be
searched can be reduced and answer sets display associated
information together.

The second improvement which can be made is to separate out
a preferred first search level.   It is necessary, however, to
consider the many objectives possible for such a selection.
Chemical information systems can be classified into two levels
as authoritative and informative. The first level by definition
is bound to be a relatively small group of compounds.
Information on these compounds can be authenticated and
information on physical properties, hazards etc. presented to
the user in a formatted and tabular form designed to assist
decision making. In many cases the data would be valuable subset
of structures useful for mathematical models, and for display in
the form of data sheets. The second level is the comprehensive
coverage suitable for general information enquiry and
browsability.

To   attain   an   authoritative   information   level   the
co-operation of the chemical profession is essential. It is at
this stage that it can begin to recover from its tarnished
image. There are many examples of co-operative ventures carried
out by chemists which have resulted in excellent and useful data
bases and publications.   The original work of Heilbron and his
associates produced the Dictionary of Organic Compounds and the
new edition has continued in the tradition of co-operation. The
Cambridge XRay Crystallographic file is another example, and a
new co-operative venture C_{13}NMR is now under way.

Finally, the use of chemical structure recording techniques can be used to produce accurate chemical and property data files in text-based systems. By establishing a chemical registry, using the logical chemical record, and encoding in WLN or a connection table form it is possible to produce a chemical dictionary which can be relied upon by non-chemists to provide all relevant data when using text based retrieval methods. It is not essential to go to structure based retrieval for such validated data as would be held at the Authoritative Search level.

CONCLUSIONS

The chemical profession is not making available its considerable expertise to the information profession to enable them to provide adequate services to both chemists and non-chemists. Lack of direction has produced services mainly designed to overcome the problem of volume for indexes and publishers but not users. Classification techniques although well applied by library staff have not been applied to chemical indexes. Failure to derive a logical compound record has led to the unique definition of the CA registry number and the resultant failure to provide necessary relationships. Finally the chemical profession has not given sufficient attention to the need to provide authenticated data not only to the scientific disciplines, but also to the layman.

There is a neglected ingredient in chemical computer systems. The chemical profession must apply more effort and better techniques to the indexing of chemicals and physical properties if it is to take advantage of the considerable computer power which is now available.

The rest of this conference will deal with successes in the hardware and software fields. I make no apology for concentrating this talk on the other necessary ingredient - accurate and logically constructed information files.

S.R. Heller and R. Potenzone, Jr. (Editors), *Computer Applications in Chemistry*
© 1983 Elsevier Science Publishers B.V., Amsterdam — Printed in The Netherlands

DIRECTIONS IN MACROMOLECULAR STRUCTURE REPRESENTATION AND DISPLAY

RICHARD J. FELDMANN
Division of Computer Research and Technology, National Institutes
of Health, Bethesda, Maryland 20205

The macromolecular surface display system developed at the
National Institutes of Health has been used to represent and solve
a number of macromolecular structure problems. This paper presents
the results from the system in the form of color images of
macromolecule surfaces.   The discussion attempts to show the
significance of each image as a member of a large class which
either has been generated or could be generated.

A computer based system like the macromolecular surface
display system consists of four components in almost equal part:
Physical scientists with problems, computer scientists with system
building expertise, hardware and (macromolecular structure) data.
If any component is weak or non-existing severe problems ensue.
The vitality of the system as a whole depends on a flow of
solvable problems. In this regard the NIH, the USA and the world
provide larger and larger pools of scientists with problems
concerning the structure of macromolecules. To restrict the use of
the system to one laboratory, institution or even country would
dramatically reduce its effectiveness, since experience shows that
there are strong interactions between seemingly random isolated
problem-solving events.   As the result of many individual
interactions between the system and the end-user physical
scientist it is clear that the macromolecular modeling system must
be perceived as a neutral non-competitive vehicle for solving
these problems. The neutral stance is achieved by leaving the
conceptualization of the problem to the physical scientist and by
striving to achieve a balance between physical and
computer/graphic experimentation. The balance expresses itself as
a distribution of display and modeling events. Since the
culmination of a project is normally the publication of the

results in a first-line journal in the discipline area of the physical scientist, the system must be capable of responding quickly to this need. Other scientists need structure representation midway in a project. Good examples of this have arisen recently where scientists needed macromolecular structure representation of families of proteins in order to explain independently immunological and nuclear magnetic resonance behavior. It has been possible to make the system respond to this need by developing a suite of programs to extend the crystallographic structure of a protein to the other members of the family.

Areas of scientific interest change in time so that the main task of the computer scientist is to sense the directions in which macromolecular structure problems will flow. Eight or ten years ago, the extraction of macromolecular structure data sets from the producing crystallographers and the purification and organization of a macromolecular data base were real problems. Now crystallographers more or less willingly contribute data sets to the recognized macromolecular structure data base. Today, structure data sets representing the interaction of proteins and nucleic acids are finally becoming available. In the past, the programs which constitute the system could handle either proteins or nucleic acids but as it turns out not both simultaneously. Considerable effort has been devoted in recent months to sweeping the required changes throughout the programs of the system.

Many of the examples which follow represent very large macromolecular assemblies. These images, the underlying forms of data manipulation and the resulting structure data sets act as a stimulus for formulating problems at this scale of macromolecular structure. In the future, it seems that more and more of molecular biology will be focused on cellular level events. Methods will have to be developed for representing the temporal and spatial distribution of structures and events at the cellular level. Fortunately, new extremely powerful, yet relatively inexpensive hardware is coming into the marketplace for achieving this type and level of structure representation.

The hardware of the system consists of an Evans and Sutherland PS-2 line drawing display and an Evans and Sutherland raster display. Both displays are hosted by a DEC PDP-11/70 computer. All of the accompanying space filling examples were produced from the monitor of the raster display. The resolution of the raster display is 512 pixels across the screen and 480 lines down the screen. Each pixel represents one byte of information enabling a color menu of 256 colors. Each color is a composition of the three basic colors (red, green and blue) in various intensities. The algorithm for drawing the shaded spheres was done by Thomas K. Porter (ref.1). The sphere algorithm is written in BLISS an implementation language for the PDP-11 but the main application programs are written in FORTRAN. The driver for the sphere program is called PROT because it manipulates the data sets for proteins as well as nucleic acids. The application program which does macromolecular modeling is called WHOLE because it was the synthesis of several earlier attempts at modeling large structures. These display and modeling programs have recently been converted to Evans and Sutherland MPS color line drawing display and the DEC VAX-11/780 computer. This configuration of hardware with a second generation Floating Point Systems array processor will be the world standard for the next five years.

The preparation of macromolecular data sets is done on a DECsystem-10 computer using programs written in SAIL. A discussion of the philosophy and implementation practice of these programs is given in an earlier review (ref. 2). The macromolecular manipulation programs have been used in the execution of two large projects; AMSOM (the Atlas of Macromolecular Structure On Microfiche, ref. 3) and TAMS (Teaching Aids for Macromolecular Structure, ref. 4). Effort is underway to translate these manipulation programs into a more portable language such as PASCAL, C, MAINSAIL or perhaps even FORTRAN.

Up until recently, macromolecular structure display and modeling has been almost totally dependent on the results of x-ray crystallography. Only in the last year or two has an approximate balance been struck between the dependence on crystallographic results and the capabilities of de-novo structure modeling. The examples which follow illustrate a number of aspects of this balance. The problem lies in the fact that it is extremely difficult to generate atomic-level detail in any consistent fashion. Only when high symmetry exists is there usually any chance of making a reasonable structure. For example, a double helix of DNA can be generated given the structure of a pair of nucleic acids. It is much more difficult to do the same thing just

starting from the chemical composition. A polymer of a protein
monomer can be generated if the packing symmetries are known but
to model the atomic-level detail is still very difficult. See
Plates 11 and 12 for a low resolution packing. Plate 10, on the
other hand, shows one assembly where we have managed to make a
reasonable atomic-level model from first-principles. Any
collection of macromolecular structure representations should
start with a statement of the 'first law of protein structure',
the alpha helix which was first postulate by Linus Pauling. Plate
1 shows two versions of an alpha helix extracted from sperm whale
myoglobin. The left helix represents all the atoms in the section.
The color code is the familiar CPK (Corey, Pauling, Kolton) color
code used in the plastic models. Red is oxygen, blue is nitrogen,
white is hydrogen and grey is carbon. The right helix shows only
the atoms of the peptide backbone plus the beta-carbon. In the
left image it is difficult to perceive the helix construction, in
the right image the helical construction is quite clear. The
modeling system provides commands for selecting structure
sections, for abstracting structure sections and for using color
to emphasize various aspects of a structure.

Plate 2 illustrates the use of color. The molecule is
cytochrome c determined by Richard Dickerson and his co-workers at
the California Institute of Technology. Cytochrome c contains a
heme which is shown in red, and many lysine amino acids which are
shown in yellow. The other portions of the macromolecule are
colored a neutral turquoise. This image shows how the lysines are
distributed over the surface of the protein and that the edge of
the heme is accessible to the solvent.

The relationship between a protein and its substrate is
illustrated in Plates 3 and 4. The molecule is hen egg white
lysozyme and its structure was determined by David Phillips and
his co-workers at Oxford University. The binding of a
polysaccharide in Plate 4 is the result of calculations by Matthew
Pincus and Harold Scheraga at Cornell University. In Plate 3 the
lysozyme is oriented so that the active site runs horizontally
across the image. The active site is a broad deep groove. The
coloring scheme used in this image is described as 'functional'
since the colors describe the function as well as the type of the
atom. The strongly charged oxygens are a deep red and the weakly
charged carbonyl oxygens are a pink. The strongly charged

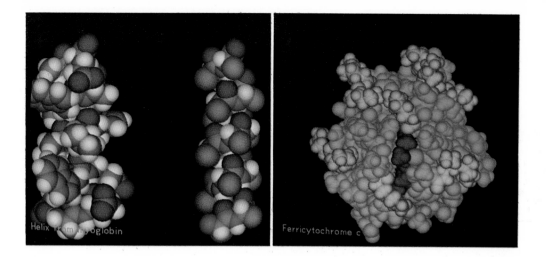

Plate 1. Helix from myoglobin.

Plate 2. Ferricytochrome c.

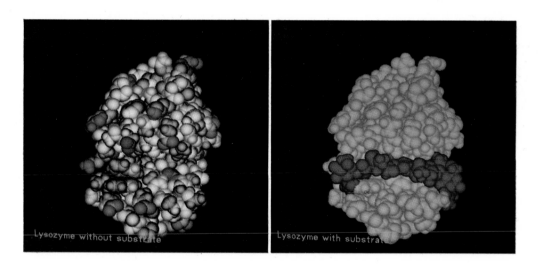

Plate 3. Lysozyme without substrate.

Plate 4. Lysozyme with substrate.

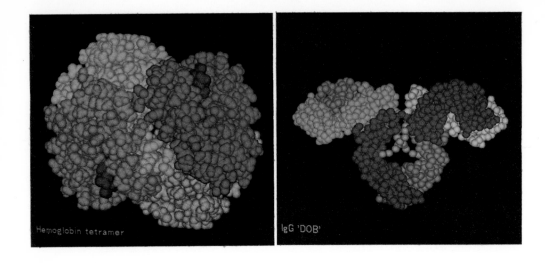

Plate 5. Hemoglobin tetramer.   Plate 6. IgG 'DOB'.

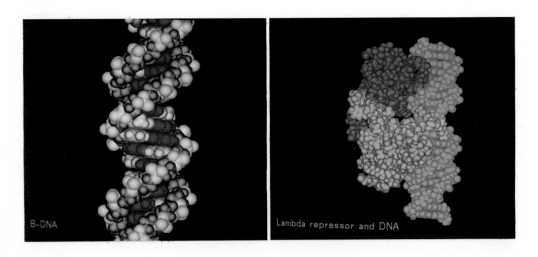

Plate 7. B-DNA.   Plate 8. Lambda repressor and DNA.

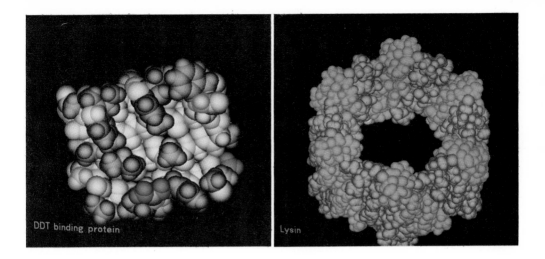

Plate 9. DDT binding protein.   Plate 10. Lysin.

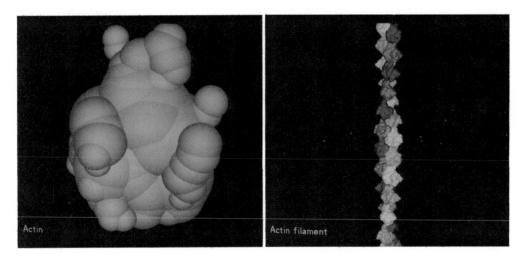

Plate 11. Actin.   Plate 12. Actin filament.

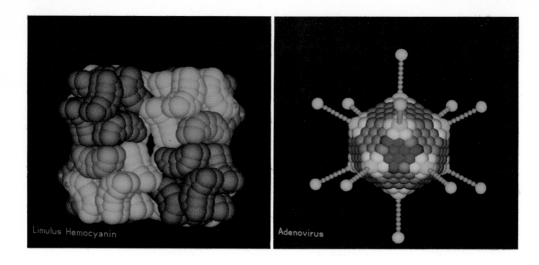

Plate 13. Limulus Hemocyanin.

Plate 14. Adenovirus.

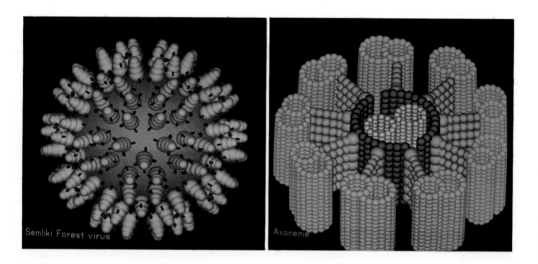

Plate 15. Semliki Forest virus.

Plate 16. Axoneme.

nitrogens are a deep blue and the amide nitrogens are a light blue. The strongly hydrophobic carbons are a dark grey and the weakly hydrophobic carbons are a light grey. The tyrosine and serine hydroxyls are white. Hydrogens take on the color of the heavy atom to which they are bonded. The functional coloring scheme is an improvement on the CPK coloring scheme used in the plastic models because the hydrogens which constitute half of the atoms in a molecule do not cause visual clutter. Plate 4 is colored to emphasize the spatial relationship between the lysozyme and the polysaccharide.

The display system is capable of representing large complexes as shown in Plate 5. The structure of hemoglobin was determined by Max Perutz and his co-workers at the MRC Laboratory of Molecular Biology in Cambridge. The tetramer of hemoglobin with hydrogens is approximately 9500 atoms. The algorithm which generates the spheres representing each atom actually calculates the shape and shading of each atom even though only twenty percent of the atoms are visible in the resulting image. This algorithm requires the atoms to be sorted as a function of where the atom begins in the vertical (i.e. Y) dimension of the screen. The implementation of this sorting feature, in effect, limits the number of atoms in a given image to about 10,000. This hemoglobin complex is colored to emphasize the relationship between the four monomers as well as the relationship of the heme to each monomer. Only two of the hemes which are colored red are visible in this orientation of the tetramer.

Plate 6 illustrates a molecule larger than the hemoglobin tetramer with fewer of the sphere primitives. The intact immunoglobulin IgG 'DOB' is represented by one sphere for each amino acid. The red and blue each represent one of the heavy chains and the green and yellow each represent one of the light chains. The grey spheres represent the associated poly saccharides. The structure of this model was determined at two different levels. The structure of the arms of the T, the so called FAB fragment, was determined at atomic resolution by David Davies and his co-workers at the National Institutes of Health and the vertical portion of the T, the so called FC fragment, was determined at atomic resolution by Robert Huber and his co-workers at the Max-Planck Institute in Munich. Using these components David Davies and Manuel Navia modeled the relationship between the

FAB arms and the FC stalk as well as the structure of the carbohydrate.

The classic B form of DNA is shown in Plate 7. The prototypical base paired structure was determined by fiber diffraction methods by Struther Arnott and his co-workers at Purdue University. The polymer shown in this image is a random sequence of base pairs designed to show all of the pairwise interactions between the four types of nucleic acids which constitute all DNAs. The coloring scheme which was designed in collaboration with Ram Sarma at SUNY Albany, emphasizes the relationship between the phosphate (white) and ribose (black) backbone of the double stranded structure and the bases (the purines are red and the pyrimidines are pink). Over several years a number of related coloring schemes have been developed to satisfy the needs of individuals and projects. Clearly, color is a very important variable in the representation and display of structure information. A program, DNARNA, exist for synthesizing double stranded structures in any form (i.e. A, B, C or Z) as DNA as well as RNA. This program is very popular with geneticists who are studying quite carefully the relationship between nucleic acid sequence and the action of proteins on such sequences.

The complex of the N-terminus of the lambda repressor and DNA is shown in Plate 8. The structure of the lambda repressor was determined by Mark Ptashne and his co-workers at Harvard University. The relationship of the repressor to B form DNA is modeled so that the total image is part crystallographic, part fiber diffraction and part modeled. This type of modeling is becoming more prevalent because it is frequently very difficult to obtain crystals of proteins complexed to DNA. The DNA is colored a neutral turquoise and the two equivalent monomers of the lambda repressor are colored blue and yellow. There are two really interesting aspects of the repressor structure, the first is the helical portion of the repressor which sits in the major groove of the DNA and causes the recognition of a specific sequence of bases and the second is the arm which reaches out to bind the repressor to the DNA. In this image the arm is easily recognized but the helix in each repressor monomer is not separately colored and hence appears as just part of the main body of the repressor protein.

Beginning with Plate 9 we leave the realm of crystallographic

truth and enter the realm of pure macromolecular modeling. In the
same way that the images of the crystallographic structures
started with a 'classic', the modeled structures start with the
first example of a protein designed for a specific function by a
human from first principles. The molecule shown in Plate 9 is a
small polypeptide of 24 amino acids designed by Bernd Gutte of the
Biochemical Institute in Zurich to bind DDT. Gutte is a peptide
synthesist who thought about the smallest architecture a peptide
could use. He focused on a very small beta sheet and designed into
it three classical beta turns. The remaining amino acids were
specified to generate a pocket for the binding of DDT on either
side of the flat anti-parallel beta sheet. The peptide has been
synthesized and has a respectable binding affinity for DDT. The
model of the structure was generated by cutting out four strands
of the main beta sheet from concanavalin A. The structure of
concanavalin A was determined by Karl Hardman when he was at the
Argonne National Laboratory. The beta sheet from concanavalin A
was used as the structural basis for the protein extension suite
of programs which is part of the molecular modeling system. The
resulting model is shown in Plate 9 in the functional coloring
scheme. The molecule appear to be rather hydrophobic which is
reasonable since it must approach the membrane where the DDT
settles and then selectively compete with the membrane for the
binding of the rather hydrophobic DDT.

Gutte's molecule compresses into one example many of the
dreams that I, for one, have had about proteins. Given a single
example, the mind fairly races ahead with extensions, variations
and developments. The original DDT binding molecule could be used
as a vehicle for testing the effects of altered amino acid
sequences on the stability of the protein. One could imagine
designing a molecule composed of two hinged beta sheets using the
amino acids to attract some substrate. The binding of the
substrate would cause the hinged beta sheets to spring together
like a venus-fly-trap. The DDT example prompts one to think again
about other classic architectural forms. Perhaps, it would be
possible to make a program to populate the amino acid backbone of
a classic architectural form to obtain a specific binding or
enzymatic activity.

Atomic resolution modeling of protein assemblies from first
principles is just becoming possible. Plate 10 shows the structure

of lysin.   It is an assembly of twelve identical proteins each of
27 amino acids.   Robert Guy at the National Institutes of Health
has analyzed the amino acid sequence and determined that the
sequence has a high propensity for alpha helix.   The helix is
strongly hydrophobic on one side and strongly hydrophilic on the
other side. Using first Labquip models and then CPK models and now
computer models we have been able to get quite a number of
isolated pieces of physical data and structure to come together.
The assembly is formed by turning the hydrophobic amino acids to
the exterior so as to form a hydrophobic interface with the
membrane.   The hydrophilic amino acids as a consequence are turned
inward.   These amino acids form salt bridges with each other thus
stabilizing the whole assembly.   By forming salt bridges the
charged groups of the hydrophilic amino acids are effectively
neutralized so that the interior hole with approximately 30
angstroms in diameter is fairly neutral.   Plate 10 shows the
twelve helices alternately colored turquiose and yellow.   The
assembly is tilted 45 degrees to show both the interior and
exterior construction.
    Atomic level modeling of protein monomers and complexes is not
always possible. In the case of the actin monomer, Uli Aebi at the
Johns Hopkins University had only very low resolution diffraction
data. He had been modeling actin at various levels by making balsa
wood models from the electron density contours. We developed a
program to transform the electron density contours to a collection
of spheres.   The resulting model of the actin monomer is shown in
Plate 11.   This model quite nicely represents the shape of the
diffraction density.   The spheres in this use represent only shape
not atomic or amino acid composition.   The advantage of having a
computer model of the monomer is evident in Plate 12.   The polymer
is simply constructed by using the helical pitch and rise which
can be obtained from the diffraction data.   The resulting polymer
shows the association between elements of the two strands of the
helix.   Approximately fifty spheres are used to represent each
monomer.   In time, we would like to represent much larger
assemblies of actin. With the present model complexity and the
hardware constrains of 10,000 spheres, we could make an assembly
of 200 protomers.   By reducing the number of spheres in the
protomer the number of protomers could be proportionally
increased. It might be possible to make an adequate representation

with only 5 spheres.

Plates 13 to 16 illustrate the structure of a variety of low resolution models of protein assemblies. Each monomer of the limulus hemocyanin shown in Plate 13 is represented by twenty spheres. The assembly of 48 protomers which was generated in collaboration with Jean and Josette Lamy of the University of Tours is composed of four assemblies of 12 protomers. Each 12-mer is composed of two 6mers. Within each 12-mer there are eight distinct proteins which are thought to be necessary to obtain a specific assembly. This example is colored to distinguish the 6mers and one of the specific proteins. The adenovirus shown in Plate 14 is icosahedral in form. The twelve spike proteins are represented by turquoise and green spheres. The proteins directly adjacent to the icosahedral spikes, the so called peri-pentonal proteins, are colored yellow. The remainder of the virus coat protein assembly is formed by attaching twenty sets of nine proteins. One of the 9-mer assemblies is colored red in this example.

The model of the semliki forest virus shown in Plate 15 was synthesized from data presented in an article in Scientific American. One graphical sphere primitive represents the structure of the entire membrane of the virus. At the other extreme of scale, seven red spheres are used to represent the structure of each of the sugars on the surface of the yellow spike proteins. A hemisphere of this virus uses about 7000 spheres.

The model of the axoneme shown in Plate 16 is the largest scale generated to date. This model indicates that the sphere primitive will probably not be useful for representing structure at the cellular level. The Lexidata Solidview system which very rapidly generates images composed of polygons will probably be the most realistic way of generating images of cellular structures.

CONCLUSIONS

The macromolecular modeling system developed at the National Institutes of Health is capable of displaying, representing and modeling a wide range of macromolecular structures. The system is a powerful, flexible collection of programs running on state of the art hardware. At the NIH the system serves as a national resource for macromolecular modeling. Export copies of the system are now working in chemical and drug companies throughout the world.

The algorithm for computer representation of the surfaces of molecules has had a profound impact on our thinking about macromolecules. Previously unintelligible structures come to life and understanding in images like those presented in this paper.

The modeling system has developed to the point where it is possible to make atomic-level models of macromolecules from first principles. Although only a few examples have been done to date, it is clear that in a short time it will be possible to make models of proteins from just the amino acid sequence.

The large-scale low-resolution modeling of assemblies is just developing. The examples shown in this paper indicate the range of scales which can currently be encompassed. These images are acting as a stimulus to the thinking of biochemists, geneticists and biologists.

REFERENCES

1 Porter, T. K, Spherical Shadiong, Computer Graphics 12 (1978) 282-285.
2 Feldmann, R. J., The design of computing systems for molecular modeling, Annual Reviews of Biophysics and Bioengineering, 5:477-510, 1976.
4 Feldmann, R. J., Atlas of Macromolecular Structure On Microfiche (AMSOM), 1977, Tracor Jitco Inc., Rockville, Md., 725pp and 650 microfiche. (110,000 pages of text and stereo graphics).
5 Feldmann, R. J., Teaching Aids for Macromolecular Structure (TAMS), 1980, Taylor Merchant Inc., New York, New York, 140pp and 116 stereo slides and student unit of 49 stereo slides.

S.R. Heller and R. Potenzone, Jr. (Editors), *Computer Applications in Chemistry*
© 1983 Elsevier Science Publishers B.V., Amsterdam — Printed in The Netherlands

SOME ASPECTS OF THE APPLICATION OF PATTERN RECOGNITION METHODS IN CHEMISTRY

K. VARMUZA

Technical University of Vienna, Institute for General Chemistry, Lehargasse 4, A-1060, Vienna, AUSTRIA

Any sense-impression and many interpretations of data in science, technology and medicine can be considered as the recognition of patterns. In a general meaning a pattern consists of a set of features which characterize some properties of an object or an event. Decisions about objects or events in a complex real-world problem are always based on a set of several features but not on single numbers (measurements). In this general sense chemistry is full of pattern recognition applications.

Strictly speaking about pattern recognition applications in chemistry means mathematical methods for the classification and interpretation of multivariate chemical data. Objects to be classified, like samples, materials or chemical compounds are characterized by sets of (measured) chemical features. A set of features is called a pattern. Spectral features, concentrations of components or a set of descriptors for the molecular structure are commonly used pattern components. A pattern can be represented by a point in a multidimensional space by using each feature as a coordinate. Pattern recognition methods look into this multidimensional pattern space.

One purpose of pattern recognition is to categorize a pattern as a member of the class to which it belongs. Class membership is connected with a property which is not directly measurable, e.g. the origin of a sample or presence of a molecular structure. Another aim of pattern recognition is to find natural clusters of pattern points which correspond to useful chemical groups (cluster analysis). Finally the location of a pattern point within a cluster can be used for semi-quantitative determinations of interesting properties, e.g. a biological activity (ref. 1).

Identification of an object by comparison of a measured pattern
with a large set of known patterns (e.g. spectral library search)
is usually not called pattern recognition. Pattern recognition is
always associated with decisions only between a few classes.

Historical Remarks

Extensive applications of pattern recognition methods in
chemistry were started with pioneering works by Isenhour, Jurs,
and Kowalski 1969-71 (refs. 2, 3). The "roaring seventies" of
pattern recognition in chemistry began with much optimism into
this approach. A great number of mathematical methods were
applied, mainly for the automatic prediction of molecular
structures from mass spectra, infrared spectra or nuclear magnetic
resonance spectra.

The next years were strongly influenced by two papers. In 1972
Kowalski and Bender (ref. 4) introduced pattern recognition
methods in a broader scope for interpreting chemical data, and
classification of materials became an important application.
However in 1973 a paper by Ting et al. (ref. 5) and discussions on
this paper made pattern recognition suspicious to many chemists.
The attempt to correlate mass spectral features with biological
activity has led to violent replies in the literature; but
criticism at this work was rather destructive. Nevertheless some
chemists felt unpleasantly by the idea, a "learning machine" may
be able to replace trained chemists. They inhibited successful
applications of this method and for a small group of chemists
pattern recognition is still a method for determining the number
of characters in the name of a substance on the basis of a
spectrum.

However pattern recognition methods and other mathematical
methods became valuable tools for processing and analysis of
chemical data sets. A specific own name of this discipline seemed
to be useful and 1974 "chemometrics" was born (ref. 6). The first
book about "Chemical Applications of Pattern Recognition" by Jurs
and Isenhour (ref. 7) summarizd the impressive pioneering work on
this topic. A program package "ARTHUR" (ref. 8) was developed at
the University of Washington in Seattle and distributed since 1975
over many laboratories in the world.

An important progress was in 1976 with the introduction of the
SIMCA method by Wold (ref. 9) into chemical and biochemical data

analysis. Pattern recognition applications in chemistry now reached a level above simple yes/no classifications.

During the next year concrete applications of pattern recognition methods dominated over proposals for new methods. Within chemometrics pattern recognition remained an important pillar and finally became an established method. A recently published book (ref. 10) contains an overview about the first 10 years of this subject and cites 364 papers with explicit applications of pattern recognition in chemistry. Today it is a necessity for chemists who are confronted with the interpretation of multivariate data to apply pattern recognition techniques. These methods are - if properly used - a simple and reliable tool for objective pre-interpretations of data.

Education

An established method needs to be taught to students. In my opinion the value of pattern recognition should not be over-estimated but the philosophy and some very basic principles of pattern recognition would be useful for any chemistry student today. No special knowledge about computers is required for a basic understanding of this method. Details of the method should be reserved to specialized courses.

Main problems with courses on pattern recognition in chemistry and generally in chemometrics might be:

1. To find suitable teachers (usually professional "pattern recognizers" from electronic departments or professional statisticians have very little success with chemistry students).

2. To find a place in the overstocked curricula; suitable courses are Analytical Chemistry and Mathematics for Chemists.

Classifiers

All human senses have the capability to recognize simultaneously a whole pattern of impressions. Furthermore the human brain is trained and capable for a fast interpretation of these patterns. Modern analytical instruments can also produce in short time large number of measurements. E.g. a resulting spectrum or chromatogram are numerical patterns which are typical for the analyzed sample. However a table of numbers is difficult for a human interpretation and the analyst suffers now under two problems instead of one: first the chemical problem, and second the numerical problem. Many chemical-analytical problems are complex and require urgently the acquisition of many data. A bottleneck is therefore often the lack of an automated processing and interpretation of the data.

Starting point for all pattern recognition applications is a relatively large (multivariate) data set (Fig. 1). Conclusions which are drawn from the data may have two different aims:

1. A first aim is to get useful information about the data structure by cluster analysis. Modelling of clusters gives relationships between measured data and not-measurable properties. The relevancy of the data (and instruments) for a given problem can be investigated by feature selection methods.

2. Another aim is a concentration of data into a classifier for future use. A classifier is a more or less simple algorithm that is able to classify new (single) samples on the basis of measurements. Many proposals for pattern classifiers in chemistry deal with binary classification problems. A binary classifier discriminates between two alternative classes and is often defined by a linear equation.

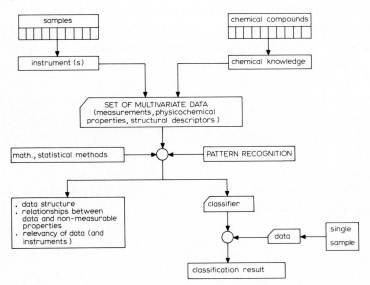

Fig. 1. Typical applications of pattern recognition in chemistry.

$$s = w_1 x_1 + w_2 x_2 + \cdots + w_d x_d$$

$x_1 \cdots x_d$ are the features (measurements) of an object (sample)
$w_1 \cdots w_d$ are (positive or negative) weighting factors (these are
the parameters of the classifier).

The sign (and value) of the scalar product s indicates the
class membership.

Classifiers can be computed from data sets with known class
memberships (supervised learning). A great variety of methods have
been used for chemical classification problems: classification by
centres of gravity, learning machine, linear regression, and
simplex optimization. An important point for the user of a
classifier is the objective evaluation of a classifier. The
following considerations are dedicated especially for spectral
classifiers.

Information win by classification

Many objections against the usefulness of pattern recognition
methods for chemical problems are legitimated because the
statistical evaluation was performed unsatisfactorily. The actual
merit of a classifier in praxis is of course also dependent on
non-objective criteria like the demands and knowledge of the user.

A useful classification is connected with some decrease of
uncertainty about a class membership. A scientific definition of
uncertainty is given by the entropy. In statistical thermodynamics
entropy S can be defined by the Boltzmann equation

$$S = k \ln W$$

W is the number of quantum states of a material system and it is
postulated that every quantum state is equally probable.

Information theory uses the term entropy in a very similar
manner. Entropy is a measure of uncertainty of signals. A signal
source that can produce n different signals has an entropy H

$$H = ld\ n \qquad (ld = {}_2\log)$$

if all n different signals are equally probable. The unit 1 bit of
entropy H is achieved for n = 2.

Let us assume an evaluation set of patterns (spectra); each of
the spectra either belong to class 1 or to class 2; the number of
patterns in each class is assumed to be equal. In terms of
information theory the uncertainty (entropy) of class membership
is therefore 1 bit. A single unknown pattern which has to be
classified can be considered as a member of such a pattern set
because before classification of the unknown equal probabilities
are usually assumed for each class. An ideal classifier would
assign each pattern of the evaluation set to the correct class and
the remaining entropy after classification would be zero; the
information win would be 1 bit which is the maximum value for a
binary classification. Actual chemical classifiers make more or
less errors. The initial uncertainty about the membership to a
certain class is therefore not completely eliminated (Fig. 2). The
difference of the entropies before and after classification is
called the transinformation R   (rate of information, information
win).

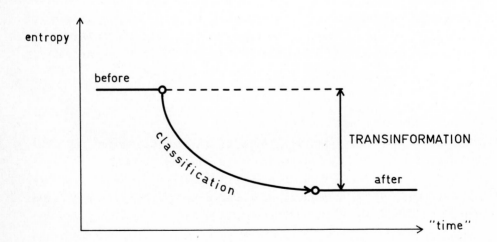

Fig. 2. Information win (transinformation) by a binary classifier.
The initial entropy (uncertainty) of class membership is reduced
by the classifier to a lower value.

The transinformation depends on the initial entropy and on the quality of the classifier. If the transinformation should be used as an evaluation criterion for classifiers, a standard uncertainty before classification must be assumed. The maximum of uncertainty is a useful and practical initial value.

The transinformation can be simply computed from two fundamental numbers, the predictive abilities P_1 and P_2 for both classes. P_1 and P_2 are percentages correctly classified patterns of class 1 and class 2, respectively (Fig.3).

$$R = f\ (P_1,\ P_2)$$

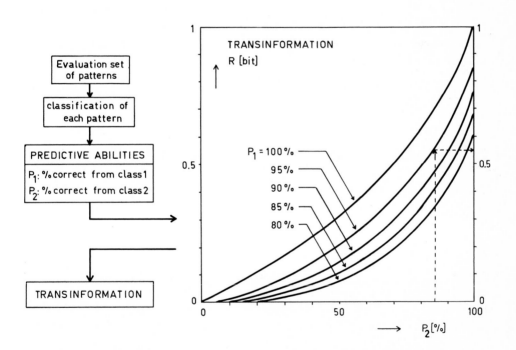

Fig. 3. Determination of information win (transinformation R) for binary classifiers as a function of predictive abilities P_1 and P_2 for both classes. Example: Classification of all members of an evaluation set give 95% (P_1) correct classifications if a pattern actually belongs to class 1, and 85% (P_2) correct classifications if a pattern actually belongs to class 2. The diagram gives a (maximum) transinformation of 0.545 bit for this classifier.

The transinformation is a single number that characterizes a classifier and can be therefore advantageously used for quality comparisons. The transinformation is measured in bit and for a binary classifier is O $\leq$ R $\leq$ 1. R=O means that there is no connection between classification and actual class membership. R=1 means that all classifications are correct (or that all classifications are wrong!; because information theory does not consider the content of signals). Fig. 3 shows that the transinformation is very sensitive in the region of "more than 80% correct classifications" which is of practical interest. Detailed mathematics are given in (refs. 10, 11).

Considerations here were restricted to simple binary classifiers with discrete answers for class assignment. However the same concept can be used for multicategory classification and for classifiers with a continuous response (refs. 10, 12).

Spectral classifiers

In the past a great deal of work in chemometrics has gone into the development of spectral classifiers. E.g. spectral classifiers have been developed as an aid for the interpretation of low resolution mass spectra of steroids. Each of these classifiers can recognize from the mass spectrum whether a certain sub-structure is present in the steroid molecule or not. The best procedures for feature generation, feature selection, preprocessing and training have been selected by using the transinformation criterion for the resulting classifiers (ref. 13). Following methods gave the best classification results for this problem: Peak heights of a steroid mass spectrum are first normalized to "local total ion current" (refs. 14, 15); that means each peak height is divided by the sum of peak heights in a window of several adjacent mass units. This normalization enlarges isolated peaks or isolated peak groups; a window width of 13 mass units yielded greatest values for the transinformation of the classifiers. A feature selection based on Fisher ratios was used for a selection of the 75 "most important" mass numbers for classification. Feature selection was done for each classification problem (sub-structure) separately. Weighting factors of linear classifiers were then computed from a set of 262 mass spectra by a linear regression analysis. This training method allows a well defined, and in some way optimal, decision plane to be achieved even for slightly overlapping clusters. The averaged

tansinformation for a set of 14 classifiers (for the recognition
of 14 different molecular structures) was 0.40 bit (ranging from
0.18 to 0.89 bit). Transinformation of 0.40 bit corresponds to an
averaged predictive ability of 85% A percentage of more than 80%
correct classifications should make this method suitable for
assisting human interpretations of steroid mass spectra.

Numerous papers deal with applications of pattern classifiers
for the automatic prediction of molecular structures from spectra.
A great advantage of this method - in comparison with library
search - is the enormous data reduction. Characteristic data about
the recognition of a certain molecular structure are extracted
from a large spectral library and a classifier is formed with only
a few parameters. Further applications of this classifier are very
simple and fast. A series of parallel classifiers should be able
to indicate for a measured spectrum absence or presence of several
sub-structures; a following computer program will combine these
results to molecular formulas. However this dream could not be
realized within pattern recognition.

Unsatisfactory results of pattern recognition methods in
spectroscopy mainly have the following reasons:
1. Primitive generation of pattern features from spectra
   (usually a simple digitization is used to obtain spectral
   features).
2. Exclusion of chemical knowledge about spectra
   interpretation.
3. The investigated classification problems are too much
   oriented on traditional chemical sub-structures, but not on
   cluster structures in the pattern space.

Applications of spectral pattern classifiers seem to be at a
dead centre in the moment. Interpretative library search systems
have been developed that are capable to give structural
information even if an unknown is not contained in the library.
New ideas will be necessary for a progress of pattern recognition
in spectroscopy. Simple pattern classifiers may be useful for the
treatment of partial problems in large artificial intelligence
programs for spectral interpretations. Furthermore the easy and
fast application make spectral classifiers useful for a
pre-selection of large spectral series as recorded with
chromatograph-spectrometer combinations.

Classification of materials

During the last years pattern recognition methods found a growing number of successful applications for the classification of materials on the basis of chemical measurements. For a great variety of materials the origin or other not directly measurable properties have been determined, e.g. for minerals, environmental samples, blood, alcoholic beverages, bacteria etc. Typical data sets contain 20 to 100 objects, each object being represented by 5 to 20 features (most often concentrations of components are used as features). These relatively small data sets (in comparison to large spectral libraries) enabled the use also of more laborious pattern recognition methods.

Cluster analysis is a standard method if an assignment of classes to patterns is not evident (unsupervised pattern recognition). The aim of these methods is to understand the relationships between either the objects or the features (ref. 16). All methods of finding clusters in a multidimensional space contain heuristic and arbitrary elements. The result of a cluster analysis strongly depends on:

1. Selection of features and mathematical preprocessing of the data.
2. Similarity criterion.
3. Clustering method.
4. Prejudices of the human interpreter about the useful number of classes.

Results of a cluster analysis therefore contain subjective aspects and should be interpreted critically. E.g. the number of the clusters must be more or less resistant against variations of the data and methods.

In a recently published paper (ref. 17) clustering techniques have been successfully applied for the elucidation and confirmation of metabolic pathways. Concentration profiles of fatty acids in the milk of several goats have been used for a clustering of fatty acids. Each group of fatty acids could be related to a specific metabolic pathway for the biosynthesis.

Because man is the best pattern recognizer known today it seems useful to include human pattern recognition capabilities into the interpretation of multivariate data. However numerical data must be transformed in a way that a human is receptive to them. An established method is the projection of the multidimensional space onto a suitable two-dimensional plane. Although much valuable information may be lost, this representation gives some insight in the complexity of a cluster structure.

Several other graphical techniques have been proposed for visual representations of multivariate data in two dimensions (ref. 18). Chemists e.g. are trained to recognize multivariate datain the form of spectra. But unconventional graphics should not be refused a priori (Fig. 4). Surprisingly, Chernoff's (ref. 19) nice proposal for a representation of a pattern by a cartoon face was not yet applied for chemical data (the features of the face would be governed by the chemical measurements). Recently analysis of multivariate medical data by this face method was published (ref. 20).

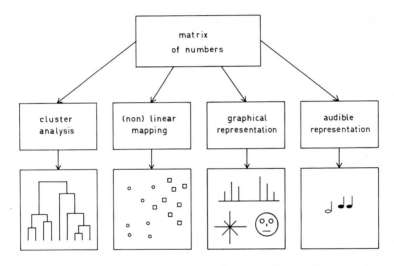

Fig. 4. Different representations of multivariate data.

Another unusual pattern recognition method uses audio representation of multivariate data. Each analytical measurement of a pattern has to be transformed into a property of a short melody which should be easily recognizable to a human. Excellent results have been reported (ref. 21) when this method was applied to the well known Obsidian data set (ref. 22). This data set contains the concentrations of several trace elements in Obsidian artifacts; these data can be used for a determination of the origin of the samples.

Although no wide applications can be expected for these exotic representations of multivariate data, there may be some chance for applications in routine supervision of production processes or complex control stations. E.g. a slight change in a well known melody may be quickly recognizable even by an unexperienced staff.

An important method for the interpretation of multivariate data is factor analysis. This method usually assumes that a single model is valid over all data. But this may be a disadvantage if one of the classes do not possess a systematic structure. An asymmetric data structure occurs if one class (e.g. biological active compounds) forms a compact cluster while the other class (inactive compounds) is scattered throughout the pattern space. Wold's SIMCA method (refs. 9, 23) is based on a separate model for each class. E.g. separate hyperboxes (or cylinders) are constructed for each class; a confidence region around the model enables the recognition of outlying pattern points (not belonging to any structured class). Furthermore SIMCA allows the quantitative estimation of so called external variables. Biological activities have been successfully estimated from the location of a pattern in relation to the model. Recently, interesting papers (refs. 24, 25) have been published about SIMCA applications for the prediction of carcinogenicity of N-nitroso compounds on the basis of their physicochemical properties.

Complex cluster structures

Modelling of classes and most other pattern recognition methods make certain assumptions about the homogeneity and shape of the clusters. Difficulties arise if a class is split into several clusters or a cluster cannot be represented by a simple geometrical model.

If the probability of each class would be known at any location in the pattern space, then an optimum classification of an unknown pattern could be made by selection of the most probable class at that point. However for a multivariate problem, the probability density curves are never known in detail for several reasons:

1. The available data are never fully representative for a classification problem.
2. The pattern space is not covered continuously by patterns of known class membership.

3.   The   probability   density   curves   cannot   be   described
     accurately by   some   parameters of a   mathematical function
     (although   this   is   assumed   in many   pattern   recognition
     methods).

4. Probability   density curves in a   multidimensional space of
    course cannot be stored point by point.

One approach for an   estimation of class-dependent probability
densities is the application of potential functions. At each point
of   the data   set (with   known class   memberships),   an "electric
charge" is   located. Each pattern is   therefore surrounded   by its
own   potential field. Different mathematical   equations   have been
used   for   the   decrease of the   potential   field with increasing
distance from the source. Superposition of the potentials from all
patterns belonging to   the same   class gives the overall potential
of that   class in   all   parts of the pattern space. The cumulative
potential can   be used as an approximation for the   probability of
that class   at a certain   point   of the pattern   space. An unknown
pattern   is classified   into   the class   which gives   rise   to the
largest cumulative potential. Advantages of this approach are that
no special assumptions   have   been made   about   the   form   of   the
probability density function and the method can also   be used   for
multicategory   classifications.   A disadvantage is   that the whole
set of known patterns is necessary for each classification and the
computing   procedure   is laborious. However   for   relatively small
data   sets with   some   dozens   of objects   -   as usual in material
classification   - and   for modern computers   this should   cause no
severe   problems. Potential methods   have been recently applied to
several   classification problems   in   chemistry and medicine (ref.
26).

    In   a   simpler   and   faster   version of the classification   by
potential functions, only a   number   of neighbors   nearest to   the
unknown are used for the calculation of the overall potential. The
well   known   K-nearest   neighbor   (KNN)   method   is   a   further
simplification.   The   K   closest   neighbors   of   the   unknown   are
selected   and   the unknown   is classified   by   a   voting procedure.
Special   voting   rules   have   been   proposed   to   overcome   the
difficulties with overlapping classes of very different size (ref.
27). Because of   its   simplicity, good performance and theoretical
foundations   the   KNN   method   serves   as   a   reference   method   in
supervised pattern recognition.

A KNN classification is very similar to a spectral library search. The distance between patterns corresponds to a dissimilarity criterion between two spectra. However a straight forward KNN classification with a large spectral library would require an enormous computation effort.

RESUME

Acquisition of chemical data from samples has been extremely facilitated by the progress in instrumental techniques. Pushing some buttons is often sufficient to produce dozens or hundreds of measurements which may be relevant for a sample or an interesting problem. Therefore it is not surprising that pattern recognition techniques - as methods for data reduction and information extraction - have become standard procedures in chemistry.

Depending on the problem in question and on the complexity of the data an assortment with pattern recognition methods is necessary. It is certainly advantageous - at least for chemistry departments at universities - to have direct access to such methods and to have some support by an expert. Unfortunately at the time no really easy to handle program package for the application of pattern recognition methods in chemistry is available. Successful applications of pattern recognition methods during the last years may perhaps encourage people to develop such programs with human interfaces.

Results obtained by the application of pattern recognition methods are not of the same reliability as results obtained by conclusions on the basis of scientific laws. One has always to consider that correlation does not necessarily imply causation. Pattern recognition methods only give a pre-interpretation which has to be followed by an investigation of the fundamental chemical relationships. Many examples in the chemical literature indicate that this investigation can often be facilitated by pattern recognition methods.

REFERENCES

1 Albano, C.; Dunn, W.J.; Edlund, U.; Johansson, E.; Norden,B.;
   Sjostrom,M.; Wold,S., Anal. chim. acta 103 (1978) 429.
2 Jurs,P.C.; Kowalski, B.R.; Isenhour,T.L, Anal.Chem. 41 (1969)
   21.
3 Isenhour,T.L.; Jurs,P.C., Anal.Chem., 43 (1971) 20A.
4 Kowalski,B.R.; Bender,C.F., J.Am.Chem.Soc. 94 (1972) 5632.
5 Ting,K.L.H.; Lee,R.C.T.; Milne,G.W.A.; Shapiro,M.B.;
   Guarino,A.M., Science 180 (1973) 417.
6 Kowalski,B.R., J.Chem.Inf.Comput.Sci. 15 (1975) 201.
7 Jurs,P.C.; Isenhour,T.L., Chemical Applications of Pattern
   Recognition, Wiley, New York (1975)
8 Duewer,D.L.; Koskinen,J.R.; Kowalski,B.R., ARTHUR (Pattern
   Recognition Program), Lab. for Chemometrics, Department of
   Chemistry, BG-10, Univ. Washington, Seattle, Wash. 98195, USA,
   1975.
9 Wold,S., Pattern Recognition 8 (1976) 127.
10 Varmuza,K., Pattern Recognition in Chemistry, Springer Verlag,
   Berlin-Heidelberg-New York (1980)
11 Varmuza,K.; Rotter,H., Advances in Mass Spectrometry 7 (1978)
   1099.
12 Varmuza,K.; Rotter,H., Advances in Mass Spectrometry, 8B (1980)
   1541.
13 Rotter,H.; Varmuza,K., Anal.chim.acta 103 (1978) 61.
14 Nageli,P., Dissertation, ETH Zurich, Switzerland (1975).
15 Rotter,H.; Varmuza,K., Anal.chim.acta 95 (1977) 25.
16 Massart,D.L.; Dijkstra,A.; Kaufman,L., Evaluation and
   Optimization of Laboratory Methods and Analytical Procedures,
   Elsevier, Amsterdam (1978).
17 Massart-Leen,A.M.; Massart,D.L., Biochem.J. 196 (1981) 611.
18 Everitt,B.S., Graphical Techniques for Multivariate Data,
   Heinemann Educational Books, London (1978).
19 Chernoff,H., J.Am.Statistic Assoc. 68 (1973) 361.
20 Honda,N.; Aida,S., Pattern Recognition 15 (1982) 231.
21 Yeung,E.S., Anal.Chem. 52 (1980) 1120.
22 Kowalski,B.R.; Schatzki,T.F.; Stross,F.H., Anal.Chem. 44 (1972)
   2176.
23 Wold,S.; Sjostrom,M., in Kowalski,B.R.(Ed.), Chemometrics,
   Theory and Application, ACS Symposium Series, No. 52, (1977)
   243.

24 Dunn,W.J.III.; Wold,S., Bioorganic Chemistry 10 (1981) 29.

25 Dunn,W.J.III.; Wold,S., J.Chem.Inf.Comput.Sci. 21 (1981) 8.

26 Coomans,D.; Massart,D.L.; Broeckaert,I.: Anal.chim. acta, 134 (1982) 139.

27 Coomans,D.; Massart,D.L: Anal.chim. acta., 136 (1982) 15.

S.R. Heller and R. Potenzone, Jr. (Editors), *Computer Applications in Chemistry*
© 1983 Elsevier Science Publishers B.V., Amsterdam — Printed in The Netherlands

KNOWLEDGE-BASE ENHANCEMENT VIA TRAINING SEQUENCE: THE EDUCATION OF SYNCHEM2

H. Gelernter, S.S. Bhagwat, D.L. Larsen, and G.A. Miller
Department of Computer Science, State University of New York at Stony Brook, Stony Brook, New York 11794

INTRODUCTION

It is a truism that the performance of a knowledge-based problem-solving system will be rigorously constrained by the depth and the scope of the information at its disposal. The task of constructing, maintaining, and expanding the knowledge base for expert problem-solving systems must therefore be seen as a crucial bridge linking the theoretical study of intelligent computer programs with the world of real problems to be solved and useful results to be attained. Indeed, it is almost fifteen years since E.A. Feigenbaum, a pioneer in the study and design of domain-specific expert consultation systems, in summarizing a report on the Stanford University Heuristic Dendral project, observed:

> Heuristic programming provided only the skeleton for problem solving .... and the computer techniques to handle the implementation. The heuristics of chemical plausibility of structures, of preliminary inference, of evaluation of the predictions, and also the zero-order and complex theories of mass spectrometry – these were all extracted from our chemist colleagues by man-machine interaction, with the program carefully guided by one of our research team. This mixed discipline for pulling out of the heads of practicing professionals the problem-solving heuristics they are using has worked far better than we had any right to expect...
> (ref.1)

Here, in a well-turned phrase, Feigenbaum has captured the
spirit of a new interdisciplinary collaboration between the
artificial intelligentsia and the community of working scientists.
But he has also underscored a major obstacle to realizing in
practice the kind of intelligent problem-solving system that seems
so clearly achievable in theory, namely the task of "pulling out
of the heads of practicing professionals the problem-solving
heuristics they are using", in order that these may be translated
into machine code to guide the heuristic search paradigms that are
emerging from this collaboration. That the process worked so well
for the Stanford group is not only a tribute to the intense
dedication and competence of its members; it was a consequence,
too, of the relatively manageable size of the knowledge base in
question.

SYNCHEM2 is a large heuristic problem-solving system which has
organic synthesis route discovery as its expert domain (ref. 2).
After more than a decade of effort devoted to the formulation and
realization of a computer environment for representing,
manipulating, traversing, and querying a problem space where the
states denote ensembles of organic molecules and the
transformations represent organic synthetic reactions taking
molecular structures into molecular structures, SYNCHEM2 is now
able to perform creditably in response to a fairly diverse range
of rather difficult problems. The knowledge base that has been
assembled to enable the system to reach its present level of
performance is, for the most part, an ad hoc and opportunistic
collection. A core of basic chemistry that any respectable
practitioner of organic synthesis could be expected to have at his
fingertips has grown by accretion to include a motley assortment
of specific, often quite powerful, and sometimes rather obscure
reactions that were added to deal with particular challenges, or
to test some feature of the overall system.

While it could be argued that the growth of SYNCHEM2's
knowledge of chemistry parallels that of a human chemist, the
absence of a systematic plan for its development has often been
painfully revealed when an impressively innovative sequence of
reaction steps proposed by the program was rendered invalid by a
single absurd or naive application of an inappropriate reaction,
or when a promising start collapsed for want of a single reaction
unknown to SYNCHEM2 at the time. It has become clear that the task

of providing a satisfactory knowledge base is not only that of assembling a suitably complete description of what is known about the problem domain, but also that of continually refining the system's understanding of what it already knows in the light of shifting foci of interest and the expansion of SYNCHEM2's problem-solving resources. Any serious attempt to design an intelligent problem-solving system intended to be useful in its target domain and not merely as a vehicle for the study of such programs must first provide the tools necessary to deal with the knowledge base. Only then can the principal task of intelligent system design be addressed in earnest.(ref. 3)

The SYNCHEM2 Knowledge Base

Two very large data bases provide SYNCHEM2 with most of the special chemistry-related information that the program needs in order to maneuver about the problem space of organic synthesis. The first of these, the Reaction Library, contains prescriptions for executing transitions between problem states; the second, the Available Compounds Library, is a collection of acceptable terminal states for the transition sequences generated through the reaction library. Together, they anchor SYNCHEM2's general problem-solving algorithm in the real world of organic chemistry.(ref. 3)

The reaction library is partitioned into chapters, each of which contains those reaction transforms (schemata) that are keyed on the synthetically significant feature (the syntheme) which characterizes that chapter. Schemata carrying more than one syntheme may be assigned to any of the corresponding chapters. Currently the reaction library holds about eight hundred active schemata distributed among twenty-eight chapters.

Each reaction has associated with its transformation pattern a collection of heuristic and physical data that enable SYNCHEM2 to qualify the use of that reaction under the conditions and chemical environment of a specific application. These data include the following:

a) Initial ad hoc estimates of the usefulness of the reaction under standard conditions. These parameters (labelled EASE, YIELD and CONFIDENCE) are used, along with other information, to compute a heuristic estimate of the relative merit of a path in the search space generated through that reaction transform.

b) Information necessary to control multiple simultaneous application of the transform when the target molecule contains multiple reaction sites.

c) Data for a series of PRE-TRANSFORM tests that check for the presence or absence of specific functionalities and structural features in the immediate target molecule to screen out inappropriate applications of that reaction before incurring the costs of executing the procedure for generating precursors.

d) Data for a series of POST-TRANSFORM tests that can be executed only after the transformation patterns have been mapped to the immediate target molecule. These procedures may, for example, examine the neighborhood of the reaction site for electronic, steric, atomic, or structural features that might enhance that instance of the reaction, or else render it invalid. The precursors generated by the schema may be checked to establish whether that instance was an intermolecular or an intramolecular application of the transform, their structural and stereochemical properties may be ascertained, and it may be determined whether invalid or unstable molecular configurations have been introduced by the transformation. SYNCHEM2 may respond to these tests by adjusting its initial estimate of the merit of that reaction application up or down, or else by rejecting the reaction step as unsatisfactory.

e) A ranking of reaction priority with respect to the other schemata in the chapter.

f) A text file containing literature references, comments, and a log recording the accession of the schema and all subsequent modifications to that reaction.

We stress the fact that the reaction library is a data structure and not a collection of programs. Every process invoked by a call to the reaction library is executed by interpreting a description of that process which is resident in the data structure. New chemistry is introduced by entering the data describing these processes, and aberrant existing chemistry is corrected by modifying these data.

The Available Compounds Library is an extremely versatile data base which has been embedded in the GRIPHOS general information retrieval processor, a hierarchical system of file definition, manipulation, and data access modules.(ref. 4) GRIPHOS provides us with an extraordinary degree of flexibility in dealing with a catalog of known compounds. For example, the amount and variety of information associated with each compound in the list may be arbitrarily reconfigured with little difficulty. If desired, the library can at some later date be expanded to include molecules with published synthesis routes, together with literature references and any additional information deemed relevant (starting materials and estimated yields and reaction merits, for example, to aid in the evaluation of a SYNCHEM2 pathway which incorporates such a known sub-path). By the same token, high-merit routes or subsequences discovered by SYNCHEM2 in the course of its explorations may also be retained in the library. With no great stretch of the imagination, one might easily interpret this last feature as an instance of machine learning in an intelligent system.

The library of available compounds currently holds about five thousand entries selected from the Aldrich, Eastman, and Frinton catalogs. While very large and complete listings of commercially available materials are now offered in computer-readable format, we decided that for the purposes of our research, an excessively rich library would hinder rather than contribute to our initial investigations of algorithm and knowledge base design. In practical applications, one would undoubtedly wish to augment the library to include all compounds that could reasonably be considered as starting materials for a total synthesis.

The Knowledge Interchange System

Traffic in intelligence travels across the borders between SYNCHEM2 and its users in both directions. The interactive interface to SYNCHEM2 and its several subsystems must support this intellectual commerce through every phase of a cycle starting with problem submission and following through with output interpretation, performance evaluation, revision and enhancement of both the knowledge base and search algorithms, then coming full circle back to the submission of a test problem. The

conversational knowledge interchange facility is expected to
provide direct and timely answers to the kinds of questions that
users invariable pose after a run, namely, "How does SYNCHEM2
propose to solve the problem?", "Why did the system select a
particular pathway at a given point during the synthesis search?",
and "Why did it fail to select, or perhaps even to generate some
reasonable alternative path at that (or some other) point in the
search?". If such information discloses deficiencies in the
knowledge base, it should then be easy for the casual user to
enter suggested additions and modifications to the knowledge base,
and for a registered experienced user to make these changes
permanent.

When fully realized, the SYNCHEM2 Knowledge Interchange System
(KIS) will be configured as an integrated system of programs which
will enable the user to pass freely among the various modes of
interaction with SYNCHEM2, its knowledge base, and the synthesis
search output. He will thus be able to pursue the logical course
of an interactive session that normally requires frequent changes
in the conversational interface without the nuisance of having to
log off one processor and onto another, and he will be spared the
concomitant danger of losing intermediate files and incomplete
results. Presently, KIS exists as a collection of discrete
modules, some in more or less polished versions, others just
recently rendered serviceable. These are currently executed as TSO
command procedures. They comprise a module for submitting
synthesis problems to SYNCHEM2, an output interpreter, a procedure
for tracing the details of the application of a specific reaction
schema to a specific target molecule, a complete data base manager
for the reaction library, and a manager for the library of
available compounds. Ultimately, these components will be bound
into a structure organized as an augmented transition network
(ATN) to create the planned unitary system.(ref. 5)

Each of the discrete interactive modules mentioned above is
sufficiently user-friendly to enable an organic chemist with
minimal computer-related experience to submit problems to the
system, examine the results, and carry through an analysis of the
output to the point where the knowledge base may be revised or
extended to reflect the results of that analysis. In routine use,
the following sequence of events might occur:

1) The problem-submission procedure is invoked. The user is asked to provide information that will be used to identify the run, and unless the run is a continuation of an earlier one, to enter a representation of the target molecule. The user is then asked whether he wishes to override any of the default synthesis search control parameters. If he responds affirmatively, he is provided with a menu of allowable adjustments, and changes are solicited. The procedure constructs a data input file for the (batch) synthesis search, and a file of job control records which may be submitted for immediate execution of the search, or else saved for later execution during a period of lower system activity, when discounted rates might be in effect.

2) Execution of the job results in the creation of an output file which summarizes the search, a status file which may be interactively queried to extract information about the search or used as input to continue the run where it left off, and a trace file (on tape) which contains a running log of the search. The tape includes the results of all heuristic tests and any debug information which may have been requested in setting up the input file for the run.

   The output summary comprises a list of all molecular structures generated during the search, a source list for the compounds found to be available, a brief abstract representation of the search tree, a detailed listing of all solutions to the problem (if any have been discovered), a detailed listing of the complete synthesis search tree, and a file of structural drawings for all (or some subset) of the molecules on the tree.

3) While the output summary and trace together contain all of the information necessary to determine what SYNCHEM2 did, and how and why the results were obtained, the completeness and detail of the former and the formidable bulk of the latter are forbidding to the casual user. Typically, the output interpreter will be invoked at this stage, instead. This processor enables the chemist to explore the search tree interactively. As the network of reactions is traversed, structural drawings are generated of the compounds at each of the visited nodes. The trace of a

terminal session with the output interpreter looks much like the notes written down by the chemist in planning a synthesis study (Appendix I). We point out, however, that even if the user should opt for interactive output interpretation in preference to wading through the full summary, he will find that the terminal session is much more effective and rewarding with a listing of the summary at his side to guide him.

4) Examination of the output invariably raises questions concerning the details of the synthesis search. While the answers may be extracted from the trace file, here too, the chemist will generally find that the desired information is more easily obtainable by interactive query. The KIS module SINGLESHOT enables him to apply any schema in the reaction library to any molecular structure in the output status file with the interactive trace turned up to any level of detail. Messages generated during this process indicate which heuristic tests have been performed, the results of the tests, and SYNCHEM2's response to the results.

5) At this point, the experienced user will wish to make modifications and additions to the reaction library in order to correct SYNCHEM2's errors and to enhance the depth and breadth of the program's grasp of organic chemistry. The KIS reaction library manager provides the chemist with a collection of interactive modules that guide him through the process of inspecting the current contents of the library, adding a new schema or modifying an existing one, validating the syntax of the changes and additions he has made, and verifying the correctness of his efforts by allowing him to apply the new or revised schema to any number of test molecular structures that he may wish to enter. Typically, a parameter or heuristic test may be revised and verified in a few minutes. Entering and verifying a new reaction may take from twenty minutes to an hour or so, depending upon the complexity of the reaction and the completeness of the heuristic tests included with the schema.

6) If SYNCHEM2 has attempted to find a synthesis for a molecule that the user knows to be available, he will probably wish to add the missing information to the available compounds library at this time.

As soon as KIS was able to support the processes described above, we had at our disposal the wherewithall to begin systematic development of the kind of knowledge base necessary to enable SYNCHEM2 to meet the intellectual standards we had set for the program. Our plan was to teach SYNCHEM2 what it needed to know in much the same way that a human student who had acquired the background necessary for comprehension could be taught how to design organic syntheses, namely through a series of problems and exercises of graded and increasing difficulty. When the program performed well we would go on with the next problem in the sequence; when SYNCHEM2 failed for want of chemical knowledge, the missing infortion would be entered into the reaction library, and the exercise would be repeated. This process, of course, would reveal deficiencies in areas other than knowledge base content. Such problems would be dealt with separately and in parallel with that of building a complete and satisfactory knowledge base for organic synthesis discovery.

The Training Sequence

"Designing Organic Syntheses", by Stuart Warren (ref. 6) was selected as the text for SYNCHEM2's course of instruction. Although there is no hint in the preface that the author had any inkling of the unconventional use his book was destined to serve, the text could have been no more suited to our needs if it had been specifically commissioned for the purpose. Indeed, when we first examined the book, we were struck with the thought that Warren might have had SYNCHEM2 in mind all along, although he insists that the book is intended for ".... a university student, a graduate, or someone with experience of organic chemistry in practice ...." Allowing for the fact that many of the important global problems of designing intelligent systems remain to be solved, SYNCHEM2 appeared to meet the requirements specified by Warren under the heading, "WHAT DO YOU NEED TO KNOW BEFORE YOU START?", namely, ".... a reasonable grounding in organic chemistry so that you are familiar with most basic organic reactions....If you are .... someone with a limited knowledge of organic reactions, you may find you need to learn some .... as you go along."

SYNCHEM2 certainly fell into the latter category, but learning

new reactions was a major objective for this study. In any event,
the book proposed to teach the student how to use these reactions
by ".... solving a series of problems of steadily increasing
difficulty .... a planned sequence of (exercises) designed to
demonstrate the use of each new concept and to test (the
student's) understanding of it. Each problem is followed by
possible solutions and full explanations...." Review and revision"
problems are provided throughout the text to allow the student to
check his progress continuously.

     SYNCHEM2's formal education began during the 1980 fall
semester. Training has continued to the present time, except for a
six month break in 1981. As of spring 1982, SYNCHEM2 had completed
somewhat more than half the material in the text. SYNCHEM2's
progress has perhaps been somewhat slower than that of a bright
graduate student, but we ask the reader to allow for the fact that
the computer program started from a rather lower base of
intellectual development than its human counterpart, and had been
primed with considerably less exposure to the methodologies of
problem-solving in the natural sciences. A brief explanation will
generally suffice to introduce a new idea to the human chemist;
SYNCHEM2 sometimes demanded the creation of a new program module
before it would absorb a novel concept.

The Training Process

     If the reader will leaf through some of his old college course
notebooks, he will readily understand that much of the detailed
record of the dialogue between SYNCHEM2 and its tutors during the
training process is not very exciting. However, the experiment did
produce many results that are instructive to examine, and some
that are of genuine chemical interest.

     An example of the former class showed up very early in the
sequence (FRAME007) when SYNCHEM2 failed to apply
(retrosynthetically) its generalized version of the Diels-Alder
reaction to 7-oxabicyclo[2.2.1]hept-5-ene-2-carboxaldehyde to
generate the precursors furan and acrolein (Fig. 1). The failure
revealed a weakness in the subgoal generation algorithm that
excludes the case where two different variable nodes in the goal
pattern must map to the same atom in the target molecule. In this
case, variables rooted at nodes a and b in the goal pattern must
both map to the ether oxygen in the target compound in order to

effect a match under the general Diels-Alder schema. Since a major
redesign of a system module (the subgraph embedding algorithm) was
indicated, correction of this problem has been deferred until the
next large-scale system upgrade is scheduled. In the meantime, the
Diels-Alder difficulty with [2.2.1] bicyclo systems and analogous
problems   in   half   a   dozen   or   so   similar   cases   have   been
conveniently managed by introducing a small number of special case
reaction schemata.

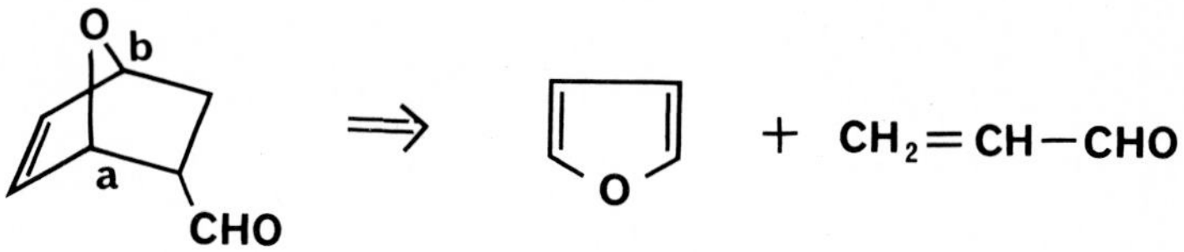

Fig. 1. FRAME007. Bicyclic Diels-Alder reaction.

A frequent occurrence during the training process is typified
by   the   result   of   FRAME029,   the   synthesis   of
1-cyclopentyl-3-phenyl-1-propanol. Here   the   problem   is that   of
dealing with reactions that may   produce isomers other   than   the
desired one.   Part of the search tree for FRAME029 is exhibited in
Figure   2, where two schemata (hydroboration of an alkene followed
by oxidation, and   reduction   of an epoxide)   each   generated   two
possible   precursors.   In   the   first   case   (hydroboration),   the
precursor that offers a degree of regiospecificity for the desired
product seems somewhat   less   satisfactory to   the   compound
complexity function than the one that would lead to a troublesome
mixture of   isomers because   an   exocyclic double bond extracts a
small penalty in perceived complexity. In the case of the epoxide
reduction,   the   spiro   precursor   would   be   reduced   to the more
substituted alcohol, while the other would again produce a mixture
of   isomers. Heuristic tests   were clearly necessary to   focus the
search on those pathways in the tree that promised the greatest
selectivity,   the   highest   yield,   and   the   fewest   byproduct
separation problems. Following   the   addition   of POST-TRANSFORM
tests to compare the degree of substitution and relative steric

hindrance at competing reaction sites for the olefin and epoxide functionalities, none of the inappropriate intermediates was accorded more than perfunctory consideration by SYNCHEM2, while the trisubstituted olefin was given favorable treatment, leading to a reasonable synthesis down that branch of the tree.

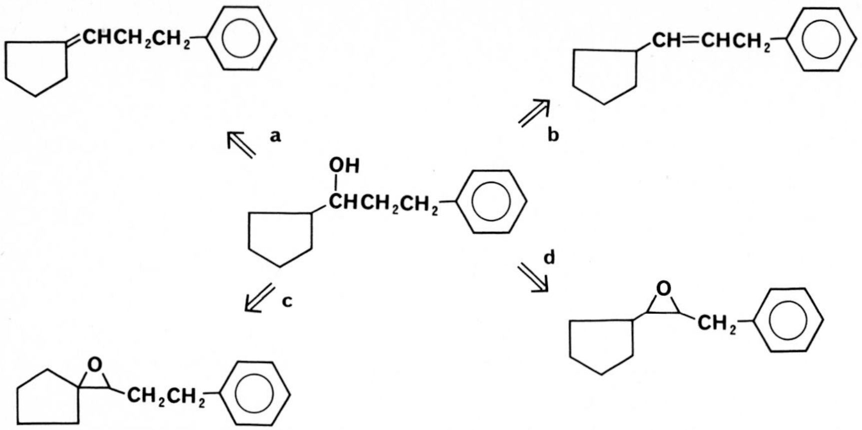

Fig. 2. FRAME029. Reactions a and b are hydroboration followed by oxidation. Reactions c and d are epoxide reductions. Pathway a is the preferred route. Pathways b and d lead to mixtures of isomers, while pathway c will produce the wrong isomer.

FRAME047, 2-methoxy-5-methyl-4'-nitrobenzophenone, presented SYNCHEM2 with the challenge of having to decide which among the several possible aromatic electrophilic substitutions are favored, and which, if any, cannot occur at all. Here, too, POST-TRANSFORM tests were added to supply comparative information concerning the estimated electronic effects associated with the substituents present on a substrate ring (Figure 3). A number of other reaction

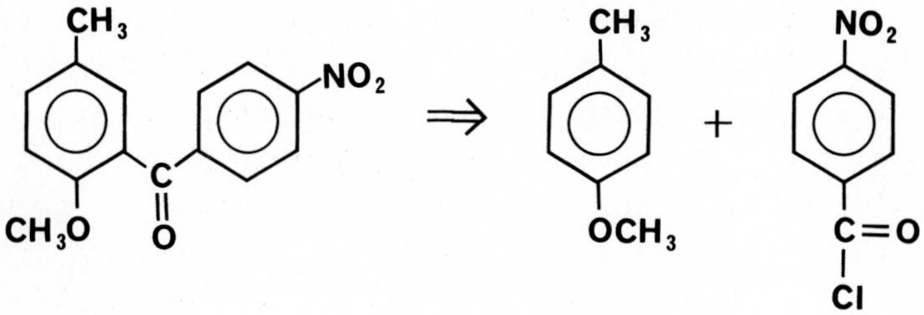

Fig. 3. FRAME047. Friedel-Crafts acylation.

schemata, particularly (but not exclusively) aromatic substitutions, have been improved significantly by the introduction of such electronic analysis. At present, when applied to aromatic systems, this uncomplicated approach is directly applicable only to aromatic monocycles. The correct analysis of ring-fused aromatics will require additional programming.

The training exercises often revealed instances where it was desirable for SYNCHEM2 to execute a series of retrosynthetic steps as a single transformation. Usually, this was because the sequence is a very common one such that the multistep schema could be expected to yield savings in search time and effort, but occasionally it was because intermediate reaction steps introduced functionalities or required procedures that had not yet been provided for in the current system. FRAME127, ethyl 4-isopropylidenecyclohexanone-2-carboxylate is an example of a problem that could be solved only after a multistep schema combining double alkylation of a malonic ester, hydrolysis, and decarboxylation into a single transformation was introduced into the reaction library (Figure 4). The new schema is a generalized synthesis of a diacid from a dihalide, and as such, is clearly a useful addition to the knowledge base. FRAME192,

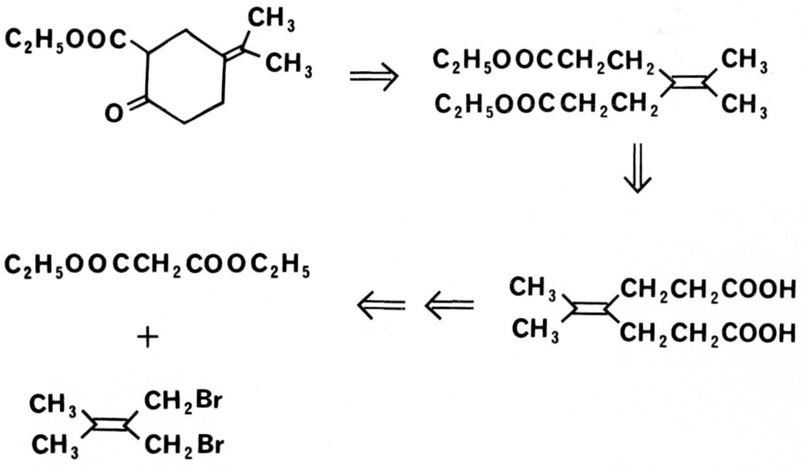

Fig. 4. FRAME127. The multistep schema, indicated by the double arrow, comprises double alkylation followed by hydrolysis and subsequent decarboxylation. The synthesis continues with esterification and then Dieckmann condensation.

1,1-bis(2-hydroxyethyl)-2-hydroxytetralin,   also   requires   a
multistep schema if SYNCHEM2 is to generate the textbook
precursor, 1,1-bis(carbethoxymethyl)-2-tetralone (Figure 5), with
the expenditure of a reasonable amount of search effort. It is not
always possible to generalize a multistep transform, however, and
the problem of FRAME192 requires such a nongeneralizable multistep
sequence. We have elected to avoid burdening the knowledge base
with massive collections of highly specific and hence rarely used
schemata, and so SYNCHEM2 has yet to produce the textbook route,
although other reasonable pathways have been discovered for this
molecule. SYNCHEM2 already contains the mechanisms for dealing
with the problem of multiple simultaneous application of schemata;
these algorithms form the basis of the MULTIPLE-MATCH procedure
that allows a single schema to be applied to multiple reaction
sites in the same molecule simultaneously. When fully implemented,
EXTENDED-MULTIPLE-MATCH will enable SYNCHEM2 to generate subgoals
that, for example, will represent precursors for the simultaneous
oxidation (or reduction) of all oxidizable (or reducible)
functionalities in the target molecule. Unfortunately, the
combinatorics of the simultaneous application of several distinct
reactions to multiple sites in an arbitrary target molecule soon
becomes unmanageable, and we have deferred extending the multiple
match mechanisms to cover such applications of more than one
distinct schema until we have invented the necessary heuristics to
control the expected combinatorial explosion.

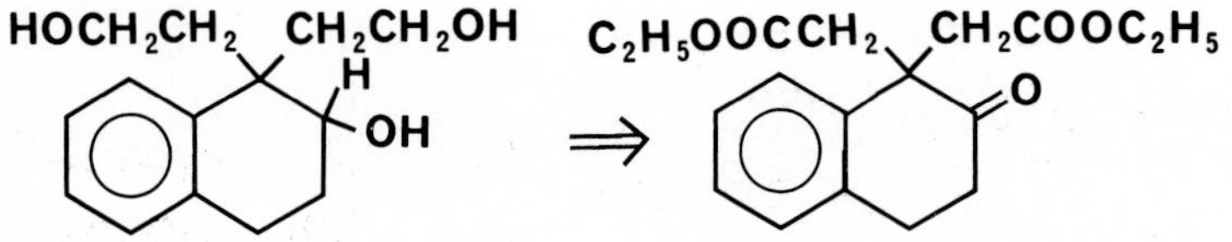

Fig. 5. FRAME192. Simultaneous reduction of esters and ketone.

The process of incremental refinement of the knowledge base is
well illustrated by the following series of events. Soon after the
set of Diels-Alder schemata was extended to encompass
[2.2.1]bicyclo systems as directed by the exercise of FRAME007, we
submitted FRAME299 (Figure 6) to SYNCHEM2 out of sequence in a

burst of bravado. The program found the textbook synthesis for
this pentacyclic diketone immediately <Appendix I>, requiring no
additions or modifications to the knowledge base at all. The final
step in the textbook retrosynthetic scheme is a Diels-Alder
reaction of cyclopentadiene with benzoquinone. In the ensuing
course of events, some of the problems submitted to SYNCHEM2 as
part of the training process were superficially candidates for
Diels-Alder chemistry. In many cases, however, the reaction was
inappropriate because the generated precursors contained at least
one active dienophile other than the one intended to react with
the diene, so that unpredictable mixtures of byproducts could
result. These unsatisfactory applications of Diels-Alder chemistry
were eliminated by introducing PRE-TRANSFORM tests to reject them
in the presence of a competing active dienophile. Still later in

Fig. 6. FRAME299. Appendix I contains a partial interpretation of
        the output for this problem.

the training sequence, SYNCHEM2 failed to find the recommended
first retrosynthetic step for FRAME222 (Figure 7), even though the
required chemistry was quite similar to that used by the program
for FRAME299, a Diels-Alder reaction with benzoquinone as the
dienophile. Analysis of the problem with the KIS module SINGLESHOT
revealed that the expected application of a Diels-Alder reaction
was prevented by the PRE-TRANSFORM tests, which had discovered a
potentially competing dienophile in the target molecule.
    The mechanism of the PRE-TRANSFORM test is obviously too
simpleminded to deal adequately with the kind of situation
described above. When alternative reaction sites are chemically
equivalent, they clearly do not interfere in the same sense that
non-equivalent sites interfere, and the desired product is
generally readily obtained through control of stoichiometry and
other reaction variables. Fortunately, existing program modules

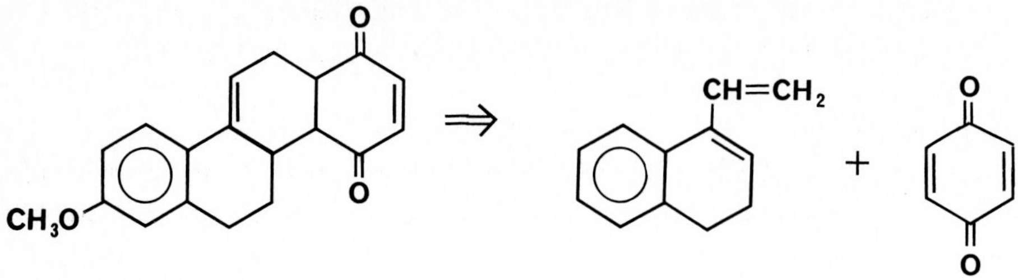

Fig. 7. FRAME222. Diels-Alder reaction.

could be easily used to define a new class of POST-TRANSFORM
tests which enable SYNCHEM2 to check potentially interfering
reaction   sites   for   constitutional   or   stereochemical
equivalence. The   discovery   of   such   alternative   reaction
sites no longer causes the routine rejection of a schema.

Reexamination of the pentacyclic diketone (FRAME299) disclosed
that if it had been submitted just a few weeks later, when the
PRE-TRANSFORM tests for active dienophiles were in place, it would
have remained unsolved at that time. Now, with the "equivalent
site" POST-TRANSFORM tests replacing the "active dienophile"
PRE-TRANSFORM tests in all Diels-Alder schemata, the desired
solution is once again discovered.

Finally, we mention a most interesting result obtained for an
insect pheromone analog (FRAME169) which was submitted to SYNCHEM2
twice, the first time in sequence about a year ago, and then again
recently as part of a check on the internal consistency of the
continuously evolving knowledge base. SYNCHEM2 was unable to solve
the problem the first time around. Even after invalid pathways
that had been generated due to inadequate screening were
eliminated by correcting the offending schemata, the parameters
for the properly applied schemata were not sufficiently well
normalized to direct the search to the long (eight step) synthesis
described in the text (and which was one of the shortest published
synthesis routes for that molecule). When the pheromone analog was
resubmitted a year later, the knowledge base had improved to such
an extent that not only was the long textbook synthesis generated,
but an original and much simpler three step route was discovered
as well (Figure 8). The new solution seems intuitively valid and
economically sound. Its feasibility under laboratory conditions is
currently being investigated.

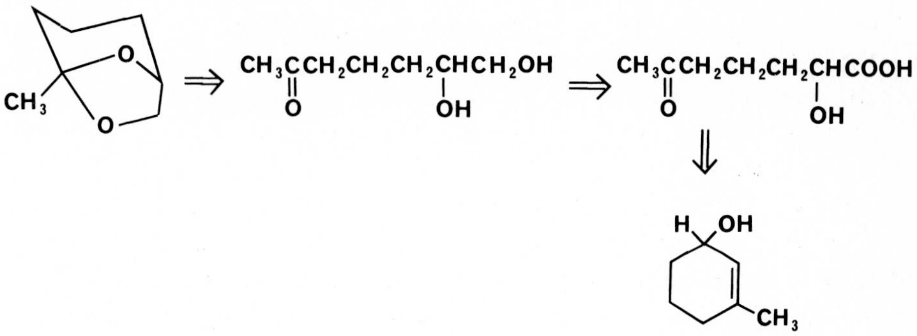

Fig. 8. FRAME169. SYNCHEM2's original three step synthesis for an insect pheromone analog. Ozonolysis followed by oxidative workup, then selective reduction of carboxylic acid, and finally, ketalization.

CONCLUSIONS

It is clear that SYNCHEM2 has a long way to go before it can regularly achieve the level of performance necessary to qualify the program for routine use in the service of organic synthetic chemistry. But it is also clear that our goal is not beyond reach.

How far beyond mere usefulness we can hope to reach with programs like SYNCHEM2 is an issue that must be broached in a different forum. The evidence accumulating in the literature has made it much more difficult to defend the position that nothing creative can ever spring from the workings of a program which is nominally deterministic. The authors obviously believe that they have only exposed the outcrop of a rich vein of possibilities. Regrettably, the analogy is more than skin deep. Digging below the surface becomes disproportionately more costly for each new nugget of progress disinterred. It is so much easier to continue harvesting the surface in artificial intelligence research than to invest the intellectual and material capital necessary to recover the buried riches. The authors have been fortunate in their choice of problem domain; the potential value of results obtained has been able to justify the costs incurred in obtaining them far longer than is usual in this kind of research, and the return promises to remain positive for some time to come.

In building SYNCHEM2's knowledge base, we have elected to follow an approach that is somewhat different from that adopted by the other active projects in computer-assisted organic synthesis design. Rather than attempt to create a fully perfected reaction schema from the outset, we have preferred instead to introduce the basic reaction as it is needed, and then to revise and elaborate the chemistry as we go along in response to performance feedback. The success of this incremental approach to knowledge base development depends critically upon the ease with which performance data may be obtained and the indicated improvements may be introduced into the knowledge base. The SYNCHEM2 Knowledge Interchange System has been designed to make such incremental knowledge base development both reasonable and feasible. We believe that the experiment described in this paper provides evidence that it works.

It is not difficult to imagine an implementation of SYNCHEM2 in which a central knowledge base is maintained and developed by a large number of expert chemists in a broadly distributed user community. Such a system configuration has obvious advantages; not only would the growth and improvement of the knowledge base be substantially accelerated, but the increased diversity of its information sources must certainly result in a knowledge base of much greater scope and depth than otherwise possible. SYNCHEM2 may be slow to learn, but what it has learned, it never forgets. It is interesting to contemplate the performance of a heuristic problem-solving system whose knowledge base has been created over the years by the combined efforts of the most talented experts in the discipline of the problem domain.

Allowing distributed access to the knowledge base, however, raises a whole new class of problems, some of which we are just beginning to understand. They concern questions of maintaining the stability and integrity of the knowledge base under conditions where multiple and possibly conflicting modifications may be introduced into the same data object by different users. Also to be considered is the quite reasonable desire of some local installations to maintain secure private knowledge bases containing possibly proprietary information, and provision of means to enable SYNCHEM2 to do the kind of chemistry at that installation which reflects the special priorities and interests of that user group. SYNCHEM2 offers a partial solution to the latter problem by allowing multiple private reaction libraries to be concatenated to the main library during a given run, but the overall issue of the distributed user community has yet to be seriously addressed.

At the risk of anticlimax, we conclude by returning to our opening paragraph. It is indeed true that the performance of a heuristic problem-solving system is constrained by the completeness of its knowledge base. The inverse of that statement, however, does not follow. Much more than knowledge about the problem domain is required before such programs can perform at truly expert levels. Pumping up the knowledge base without providing the tactical and strategic support with which to use that knowledge innovatively will make the system little more than an untiring drudge - an idiot savant, so to speak. SYNCHEM2 presently has at its disposal an effective array of search tactics to guide its traversal of the synthesis tree, and some rudimentary strategies to introduce a small measure of global planning in its overall approach to a problem. But while strategic goals can sometimes be achieved by the adroit use of tactics, SYNCHEM2 will not become the kind of intelligent system we seek to create until its capacity for strategic planning has been considerably enhanced. It seems clear that it is here that we will encounter our next major obstacles. Fortunately, SYNCHEM2 is now sufficiently reliable and stable to enable us to attack those obstacles without fear of being tripped up by our tools.

ACKNOWLEDGEMENTS

This work has been supported in part by the National Science Foundation under Research Grant No. MCS-7816786. Additional research support has been provided by the Eastman Kodak Company and by Ciba-Geigy AG. We gratefully acknowledge the assistance of the Chemical Information Systems project of the Environmental Protection Agency, MIDSD, in making available to us the computing resources that have been essential to our progress.

REFERENCES

1 E.A. Feigenbaum; "The Search for Generality", Proc. IFIPS
  Congress, 1968, North-Holland Pub., Amsterdam (1969).
2 H.L. Gelernter, A.F. Sanders, D.L. Larsen, K.K. Agarwal, R.H.
  Boivie, G.A. Spritzer, and J.E. Searleman; "Empirical
  Explorations of SYNCHEM", Science, Vol. 197, (1977) 1041.
3 Among the first to recognize the importance of "knowledge
  engineering" as a crucial element in the design of intelligent
  systems, it is not surprising that the Stanford group was among
  the first to consider the problem of providing suitable tools
  for knowledge base construction and maintenance. See R. Davis,
  "Meta-level Knowledge: Overview and Applications", Proc. 5th
  Int. Joint Conf. on Artificial Intelligence, Carnegie-Mellon
  Univ., Pittsburgh, Pa. (1977).
4 This assertion is not strictly true. As a concession to the
  interests of directness and efficiency, many of the algorithms
  have been specialized to the structures and manipulations of
  organic chemistry. Also, the function that computes the
  "synthetic complexity" of each precursor generated during the
  synthesis search (i.e., the prediction of how difficult it will
  be for the program to find a synthesis for that compound) is the
  repository of a significant measure of chemical judgement,
  reflecting as it does the combined synthesis-discovery instincts
  and experience of the several organic chemists who have
  contributed to its formulation. Nevertheless, in a world where
  the search for elegance may be permitted to supersede the
  necessity for results, the chemistry-specific modules would be
  generalized algorithms that adapted to the exigencies of the
  problem domain by interpreting the descriptions of structures
  and manipulations resident in the knowledge base. A generalized
  prediction function would then create its own parameters, both
  by synopsizing the reaction library to summarize what was
  reasonable and feasible with the chemistry currently at the
  program's disposal, and by abstracting the catalog of available
  compounds to guide the search towards those starting materials
  most readily obtainable. See (ref. 1) for a thoughtful
  discussion of this issue.

5 The GRIPHOS system was developed by J. Heller (SUNY at Stony Brook) to manage large and complex humanities oriented data archives. A complete description of the system may be found in D. Vance; "Manual for the Museum Computer Network Application of GRIPHOS", SUNY Stony Brook Spectra Pub., Cen. Contemp. Arts and Let.(1976).

6 D.J. Berndt; "A Transition Network Based User Interface", Master's Thesis, SUNY at Stony Brook (1982).

7 S. Warren; "Designing Organic Syntheses, A Programmed Introduction to the Synthon Approach", Wiley (1978).

Appendix

Output interpretation for
pentacyclic diketone, FRAME299

```
ENTER INITIALS USED FOR STATUS FILE TO BE EXAMINED.
>HLG
ENTER COMPOUND NAME (EIGHT LETTER MNEMONIC).
>FRAME299
ENTER RUN NUMBER.
>1
ENTER SEGMENT NUMBER.
>1
THE FILE 'CN.EPAHGL.WBRG.HLG.FRAME299.R1S1.STATUS2' HAS BEEN
ALLOCATED.

*** SYNCHEM OUTPUT INTERPRETER - VERSION 1.3 (10-JULY 1981) ***

LOADING STATUS FILE.
CREATE HEADER INFORMATION:

LAST        COMPOUND NODE               179
            SUBGOAL NODE               -162
            SYMBOL TABLE ENTRY          164
            SLING STORAGE RECORD          2
            ACTIVE PAGE NUMBER            0
            PAGE IN DIRECTORY            10
            ACTIVE RECORD                 0
            RECORD IN FILE               61
            EXPANDED COMPOUND           150

TOTAL COMPOUNDS EXPANDED    10
THE DIRECTORY:
   #  AGE               KEY        TYPE        DIRTY

SET DEBUG LEVEL   (0,1,2).
:
0
```

```
SET DRAWING HYDROGEN LEVEL:
    0 - NO HYDROGENS ON CARBONS
    1 - ONLY THOSE HYDROGENS THAT APPEAR IN CANNONICAL SLING
    2 - ALL HYDROGENS
:
0

SELECT TERMINAL TYPE: (1) LA120    (2) ONTEL    (3) OTHER
:
3

COMPOUND    1 IS SOLVED.
THERE ARE 12 WAYS OF MAKING THIS COMPOUND.

-8  -9  -5  -13  -2  -11  -10  -12  -6  -3  -4  -7
 D   S   D

SELECTION OPTIONS:
  S - PURSUE NEXT SUBGOAL (ENTER SUBGOAL # ON SAME LINE)
  R - RETURN TO PARENT SUBGOAL
  X - END SESSION
:
S9

CURRENT PATH:
    1
```

```
          C----C
         /|    |\                          C==C
        / |    | \                         |   \
   O====C  |    |  C====O                   |    \
        \ C    C |                          C--C--C
         \|    | |                           \    |
          |    | |           ---\             \   |
          |\   | |               \            C--C
          | C-----C          ---/            /    \
          |/   | /           ---/       O=C        C=O
          C    |/                         \    /
           \   C                           C==C
            \ /
             C

            1                                9
          SOLVED                           SOLVED
```

```
      VIA PHOTODIMERIZATION OF SOME OLEFINS
             (CHAPTER 29, SCHEMA    1)    MERIT    62
             EASE:  90    YIELD:    95    CONFIDENCE:   100

SELECTION OPTIONS:
  C - PURSUE NEXT SUBGOAL (ENTER SUBGOAL # ON SAME LINE)
  R - RETURN TO PARENT SUBGOAL
  X - END SESSION
:
C9
```

COMPOUND   9 IS SOLVED.
THERE ARE   20 WAYS OF MAKING THIS COMPOUND.

```
-32  -21  -16  -14  -18  -19  -23  -24  -25  -20  -33  -17  -26
  S         D         D

-15  -22  -29  -31  -30  -27  -28
```

OPTIONS: S (PURSUE SUBGOAL), R (RETURN), X (EXIT).
:
S32

CURRENT PATH:
   1=>   9

```
 C==C
 |   \
 |    \                          C=C            C
 C--C--C        ---\            /   \           = \
  \   |            \      O=C        C=O  + C    C
   \  |            /        \   /          \   =
    C==C         ---/         C=C          C-C
   /    \
 O=C     C=O
   \   /
    C==C
```

```
      9                        33             32
   SOLVED                   AVAILABLE      AVAILABLE
```

      VIA DIELS ALDER REACTION OF CYCLOPENTADIENE
        (CHAPTER   30, SCHEMA   2)   MERIT   88
         EASE: 90   YIELD: 80   CONFIDENCE: 100

OPTIONS: C (PURSUE COMPOUND), R (RETURN), X (EXIT).
:
R

COMPOUND   9 IS SOLVED.
THERE ARE 20 WAYS OF MAKING THIS COMPOUND.

```
-32  -21  -16  -14  -18  -19  -23  -24  -25  -20  -33  -17  -26
  S         D         D

-15  -22  -29  -31  -30  -27  -28
```

OPTIONS: X (PURSUE SUBGOAL), R (RETURN), X (EXIT).
:
R

```
COMPOOUND    1 IS SOLVED.
THERE ARE 12 WAYS OF MAKING THIS COMPOUND.

-8   -9   -5   -13   -2   -11   -10   -12   -6   -3   -4   -7
 D    S    D
```

```
OPTION: S (PURSUE SUBGOAL), R (RETURN), X (EXIT).
:
S8
```

```
CURRENT PATH:
    1
```

```
              C----C
             / |    |\
            /  |    | \
    O====C  |    |  C====O            C
         \  C----C  |                / \
          \|  |   | |              C-C     C-C
           |  |   | |              =  |     |  =
           |\ |   | |              C   C---C   C
           | C====C |               \ /       \ /
           |/   | /                  C         C
           C    |/                   =         =
            \   C                     O         O
             \ /
              C

              1                            8
           SOLVED                       DEVELOPED
```

```
        VIA PHOTODIMERIZATION OF SOME OLEFINS
        (CHAPTER 29, SHEMA    1)      MERIT    62
          EASE:    90      YIELD:    95    CONFIDENCE:   100
```

```
OPTIONS: C (PURSUE COMPOUND), R (RETURN), X (EXIT).
:
R
```

```
COMPOUND    1 IS SOLVED.
THERE ARE 12 WAYS OF MAKING THIS COMPOUND.
   -8   -9   -5   -13   -2   -11   -10   -12   -6   -3   -4   -7
    D    S    D
```

OPTIONS: S (PURSUE SUBGOAL), R (RETURN), X (EXIT).
:
S5

CURREENT PATH:
    1

```
        C----C
       / |    |\
      / |    | \
O====C  |    |  C====O
      \ C----C  |
       \|    |  |
        |    |  |
        |\   |  |
        | C=====C
        |/   | /
        C    |/
         \   C
          \ /
           C
```
```
            O
            =
            C              O       O
           / \             =       =
         BR-C   C---C--C--C-C
           \   |  /               \
            C-C-C                  O
             \  |                   \
              C-C                    C
                |
               BR
```

           1   5
        SOLVED   DEVELOPED

      VIA DOUBLE ALKYLATION, ACETOACETIC ESTER RING SYNTHESIS
        (CHAPTER 2, SHEMA   59)   MERIT   61
         EASE: 80   YIELD: 90   CONFIDENCE: 100

OPTIONS: C (PURSUE COMPOUND), R (RETURN), X (EXIT).
:
X

*** SYNCHEM OUTPUT INTERPRETER SESSION ENDED ***
READY

S.R. Heller and R. Potenzone, Jr. (Editors), *Computer Applications in Chemistry*
© 1983 Elsevier Science Publishers B.V., Amsterdam — Printed in The Netherlands

CHEMICAL INFORMATION SYSTEMS

G. W. A. MILNE

Information Technology Branch, Division of Cancer Treatment, National Cancer Institute, Bethesda, MD, 20205

SUMMARY

Some of the current problem areas in chemical information systems are discussed in some detail in this paper. The discussion is couched in terms of the National Cancer Institute's Drug Information system and reference is made also to other systems such as the NIH/EPA Chemical Information System and the DARC/Questel Substructure Search System.

INTRODUCTION - THE DRUG INFORMATION SYSTEM

The Division of Cancer Treatment of the National Cancer Institute is responsible for the routine screening of chemical substances for anti-cancer activity. Since this program began in 1955, it has examined approximately 500,000 materials for activity and is currently screening about 10,000 chemicals per year. As a result of this activity, there has been developed a large database of chemical and biological information and this is the basis of the Drug Information System (DIS).

A schematic illustration of the NCI drug development process is shown in Figure 1. The relevant literature is routinely reviewed and each year, of the roughly 450,000 new chemicals reported, about 40,000 are identified as being of potential interest to the program. Some 10,000 of these are actually acquired and tested against P388 leukemia in mice. Approximately 500 positives are found in this step and half of these are selected in a committee review for further testing in the "Tumor Panel", a group of several solid tumors in mice. Only 25 compounds meet the criteria for activity in the Tumor Panel and 15 of these are designated for continued development. Typically, only 8 of the 15 can be adequately formulated for administration to human patients and these are then submitted to a standard toxicology study. All the data on these compounds are finally submitted to the Food and Drug Administration as a request for investigational new drug status. Once this request is granted, the drug may be used in a clinical situation and responsibility for it passes to another program within NCI.

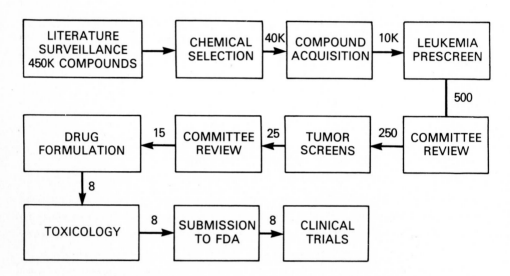

Fig. 1. Flow of Chemicals Through the NCI Drug Development System.

This entire process is moving quite steadily. Input of 10,000 candidates and production of 8 potentially useful drugs is a fairly stable annual turnover, although it can take between 6 and 10 years for any individual chemical to proceed all the way through the system. The annual cost of the entire program is on the order of $40M, which suggests a cost of $5M per candidate drug.

Much of the data management that is required by this activity is routine and will be touched on only briefly here. Some particular areas however present interesting challenges in the area of information technology and these will be discussed more fully below.

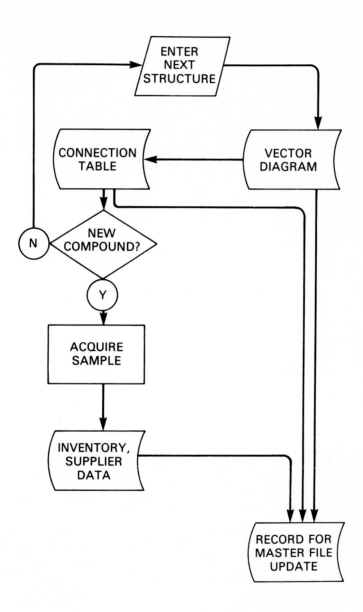

Fig. 2. Addition of Chemicals to the NCI Drug Development System.

ADDITION OF NEW CHEMICALS TO THE DIS

Once a chemical has been selected for testing, its structure
and associated information are added to a "pre-input file" which
is periodically folded into the main database. The means by which
this step is accomplished are shown in Figure 2. The structure is
entered via a graphics terminal as a set of vectors. These
describe the structure fully and esthetically. Stereochemisty is
implicit, the diagram is oriented correctly, and collisions are
eliminated.

When a satisfactory structure has been entered, a copy of the
vector set is saved and the vectors are then used to derive a
connection table for the molecule. The connection table is used in
a search of both the main database as well as the pre-input file
to establish that it is in fact a new compound, and provided that
this condition is met, the records are added to the pre-input file
and a sample of the compound is acquired. Data concerning the
supplier, material inventory and so on are added and then the
entire record is stored until the next update of the master
file.

That a compound is not already in the master file is a
necessary but not sufficient condition for its acquisition. Before
efforts are made to acquire the compound, it is tested in a model
system for its potential in three areas: activity, toxicity and
novelty. This model, known as the Hodes' model, is depicted in
Figure 3 and is based upon the statistics that have been derived
from studies of the relationships in the DIS between structure and
properties. Given a large data set, the model decomposes every
structure into a set of sub-structural fragments. It then examines
the many relationships between fragments and each of the three
properties and derives a probabilistic relationship between
presence of a fragment in a molecule and the activity, toxicity or
structural novelty of the molecule with respect to the remainder
of the database. Structures which score well in all three
categories are automatically selected for testing; low scorers are
automatically rejected and those that are intermediate are
disposed of by manual review. The model has been successful in
that it has allowed the program to decrease from 14,000 to 10,000
the number of compounds entering the system while maintaining at
500 the number of chemicals that pass the leukemia prescreen. This
leads to considerable savings and it is hoped that as the model is

further refined, these savings will increase.

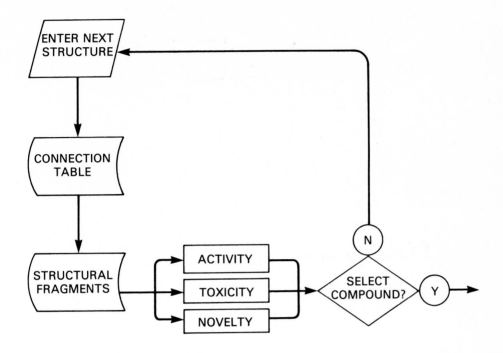

Fig. 3. Hodes' Model for the Development of Selection Criteria.

SUBSTRUCTURE SEARCHING IN THE DIS

The substructure searching capability of the DIS is modelled upon that in the SANSS component of the NIH/EPA Chemical Information System (CIS). In the example shown in Figure 4, a screen search is first carried out, for all structures having the phosphorus centered fragment shown. Only 15 structures from the SANSS database of 210,000 respond and when they are examined upon an atom-by-atom basis, only 9 survive and these are stored in temporary file #2. The first drop, shown in Figure 5, is the well-known anti-cancer drug, thiotepa, whose Chemical Abstracts Service Registry Number (CAS RN) is 52-24-4. This output provides further references to this compound; there are two online (CIS) references, and a total of 13 external files cited as carrying data on this compound. There follows the molecular and structural formulas and finally, 41 names and synonyms for the compound.

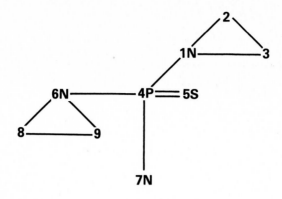

Option? FPROBE 4

Fragment:

Required occurrences for hit : 1
This fragment occurs in   15 compounds

File = 1, 15 compounds contain this fragment

Option? SUBSS 1

Doing sub-structure search
Type E to Exit

File = 2   Successful sub structures =   9

SCREEN AND ATOM-BY-ATOM SEARCH

Fig. 4. Substructure Search in the Chemical Information System.

DISPLAY OF RESULTS FROM SUBSTRUCTURE SEARCH

Option? <u>SSHOW 2</u>

How many (E to Exit)? <u>1</u>

Entry 1 CAS RN 52-24-4

Online references:
  6 — CIS, Cambridge X-Ray Crystallography: 52-24-4.01
  32 — NIOSH/CIS, RTECS: SZ2975000

Offline references:
  7 — Merck Index: 9348
  15 — CPSC, CHEMRIC
  31 — PHS-149 Carcinogenic Activity: D1423, G785
  35 — ORNL, EMIC
  36 — ORNL, ETIC
  69 — USAN & USP Dictionary of Drug Names
  72 — IARC, Monographs (Carcinogenicity Reviews)
  74 — EPA, Expanded Potential Industrial Car. & Mut.
  77 — NLM, CHEMLINE: TOXLINE, TDB, TOXBACK
  83 — USAF, Defense Pest Mgt. Information Analysis Cntr.
  88 — NCI/NTP, Chemicals on Standard Protocol: C01649
  91 — IARC, Bulletin on Chemicals Tested for Carcin.
  116 — Catalog of Teratogenic Agents: 773

C6H12N3PS

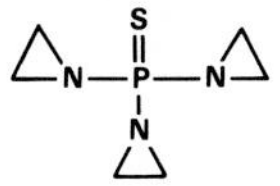

Aziridine, 1,1',1''-phosphinothioylidynetris- (9CI)
Phosphine sulfide, tris(1-aziridinyl)- (8CI)
Girostan
N,N',N''-Tri-1,2-ethanediylphosphorothioic triamide
N,N',N''-Tri-1,2-ethanediylthiophosphoramide
N,N',N''-Triethylenephosphorothioic triamide
N,N',N''-Triethylenethiophosphamide
N,N',N''-Triethylenethiophosphoramide
NCI-C01649
NSC 6396
Oncotepa
Oncotiotepa
Phosphorothioic acid triethylenetriamide
Phosphorothioic triamide, N,N',N''-tri-1,2-ethanediyl-
Phosphorothioic triamide, N,N',N''-triethylene
Tespa
Tespamin
Tespamine
Thio-Tep
Thio-Tepa
Thiofozil
Thiophosphamide
Thiophosphoramide, N,N',N''-triethylene-
Thiophosphoramide, N,N',N''-tri-1,2-ethanediyl-
Thiotef
Thiotepa
Thiotriethylenephosphoramide
Tifosyl
Tio-tef
Tiofosfamid
Tiofosyl
Tiofozil
Tri(ethyleneimino)thiophosphoramide
Triaziridinylphosphine sulfide
Triethylenethiophosphoramide
Triethylenethiophosphorotriamide
Tris(ethylenimino)thiophosphate
Tris(1-aziridinyl)phosphine sulfide
TSPA
WLN: T3NTJ APS&- AT3NTJ&- AT3NTJ
1,1',1''-Phosphinothioylidynetrisaziridine

Fig. 5. CIS SANSS Record for Thiotepa (CAS RN 52-24-4).

CAS RN 76-96-0 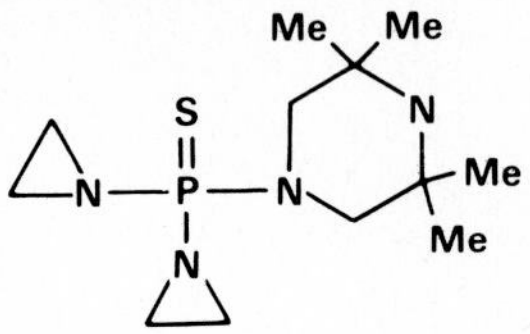   CAS RN 41657-29-8

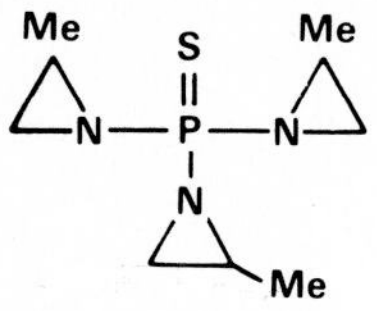

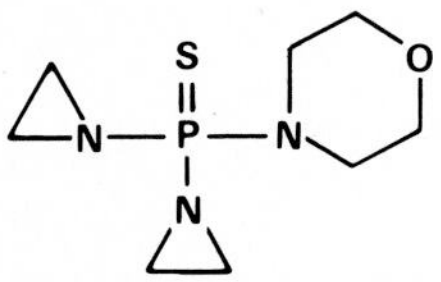

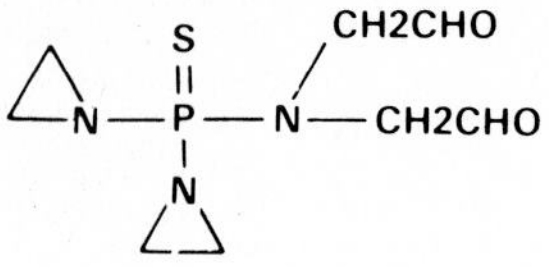

CAS RN 2168-68-5   CAS RN 41657-30-1

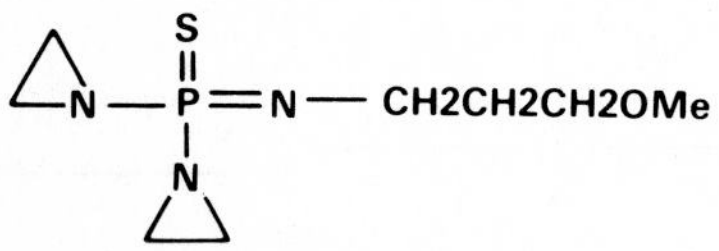

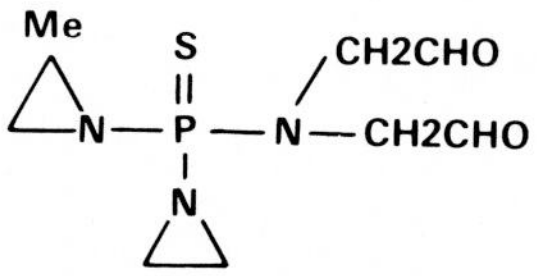

CAS RN 13687-10-0   CAS RN 41657-31-2

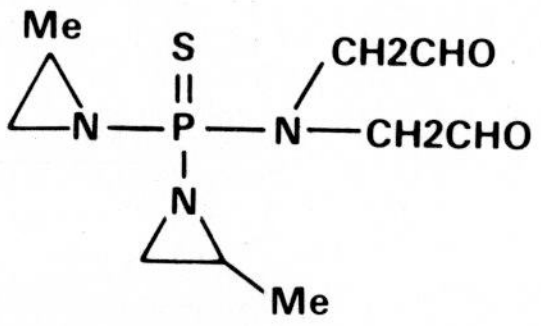

CAS RN 33683-34-0   CAS RN 64332-89-4

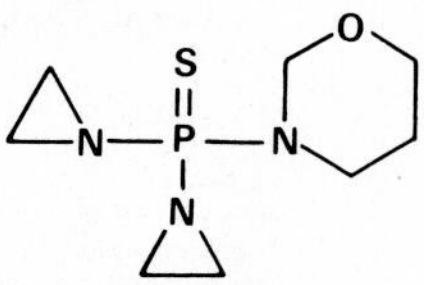

DISPLAY OF RESULTS OF SUBSTRUCTURE SEARCH

Fig. 6. Compounds Retrieved by a Substructure Search in the CIS.

Option? RING 3
Option? RING 3
Option? RING 3
Option? CHAIN 2
Option? SATOM 1 4 7
Specify element symbol = N
Option? SATOM 10
Specify element symbol = P
Option? SATOM 11
Specify element symbol = S
Option? ABOND 1 10 4 10 7 10
Option? SBOND 4 5 5 6 4 6
Bond type (H for Help) = RS
Option? SBOND 7 8 8 9 7 9
Bond type (H for Help) = RS
Option? SBOND 1 10 4 10 7 10
Bond type (H for Help) = CS
Option? SBOND 10 11
Bond type (H for Help) = CD
Option? D

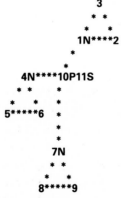

Option? IDENT
File 1, This structure is contained in   1 compounds.

Fig. 7. Full Structure Search in the Chemical Information System.

In Figure 6, there is given the CAS RNs and structures for the other 8 structures which were found by the substructure search. These all contain the central phosphorus, bonded to three nitrogens and one sulfur, but beyond that, they have only the required substructure of two ethyleneimine rings.

A full structure search is shown in Figure 7. Here the entire structure for thiotepa is entered; the search is carried out and the result is stored in temporary file #1. This is the program which is used during compound selection to determine whether or not a structure is already in the database. Major modification of this program was in fact necessary, because, as can be seen from the Figure, the structure building is quite tedious and also fails to provide the vector set which is required by the CIS.

STRUCTURE INPUT

There are two structure input programs in general use today and both are designed to produce only connection tables. Thus the SANSS structure input program, which has been adopted essentially unchanged for use in CAS ONLINE, and the Darc/Questel program both require input upon a linear atom-by-atom basis. Both programs have some short cuts, but neither make any significant use of geometric information.

The program used by the DIS on the other hand, uses a modern CRT with programmable keys. These are used to develop, with a single keystroke, a single bond at a desired angle to the horizontal and terminating in the required atoms. Other keys can be programmed to provide rings of various sizes and even entire sub-structures upon a single command. All bonds and atoms can be moved to any position on the screen and the program is thus able to produce,in vector form, structure diagrams which meet all the criteria that a user may apply. The program is very easy to use and is economical in terms of both time and keystrokes. A comparison between this structure entry program, the SANSS program and the DARC program is given in Figure 8. As can be seen from the Figure, the DIS program is between 3 and 8 times faster than the others and it produces structure diagrams which are far closer to the conventional drawings of the molecules in question.

INPUT OF COMMITTEE DECISIONS INTO THE DIS

As can be seen from Figure 1, a considerable number of

decisions that effect the DIS database are made by Committees. Such decisions result primarily in dropping of chemicals from consideration for further testing; several hundred chemicals are reviewed in this way every month. Recording of these decisions and subsequent modification of the database is a continuing task which requires substantial amounts of manpower.

In an effort to streamline this process, the DIS is experimenting with the use of a projector which behaves as a slave to an online terminal. Every chemical that is considered in committee is the subject of an interactive session lasting a minute or less. The committee sees every question on a screen and as it provides an answer to each question, the answer also appears on the screen. When the entire transaction for a chemical is complete, the computer is so informed, the master file is updated immediately, and the next candidate is called forth - both on the terminal and on the screen. As a result, the minutes of a meeting are complete at the moment the meeting ends - and can in fact be on a person's desk before that person can get back there - and there is no longer any need for the labor intensive step of preparing minutes.

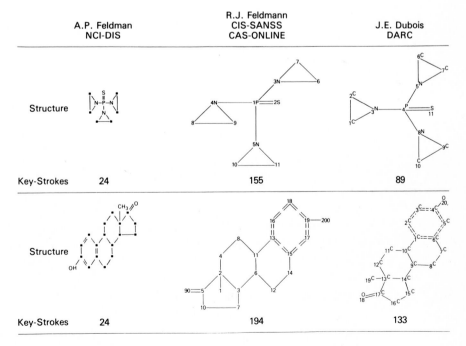

Fig. 8. Different Programs for Structure Generation.

BIOLOGY DATA IN THE DIS

The primary identifier for a chemical within the DIS is the so-called "NSC Number". This number allows the retrieval for that compound, of any of the data shown in Figure 9. In this way, not only is the chemical structure retrievable, but all the data that have been developed on a chemical may also be retrieved. This includes inventory and source information, and biological, toxicological and clinical data. Registration of chemicals for the DIS has for a number of years been carried out by CAS, under contract to NCI, and a side-product of this is that almost every NSC Number is associated with a specific CAS RN. The latter is, of course the center of a larger world of information on chemicals, as is shown in Figure 10. It is through the CAS RN that other, non-NCI data on chemicals can be retrieved, as needed.

The biology data of the DIS is organized in some 200 fields. All of these may be invoked in retrieval commands, but only 15 of them, shown in Figure 11, may be used in searching. The chemistry files can be searched at the same time as the biology files and a typical retrieval may be seen in Figure 12. Here the search is simply for a particular NSC Number, but the subsequent data retrieval is for a series of user-defined types of data on that compound. The user requested, via the FORMAT command, the structure, nomenclature, NSC Number, CAS RN, inventory and supplier(s) for that compound and these are retrieved and printed, as shown. The generation of a report of this sort, including a structure diagram if necessary, is a very important capability in the DIS and is discussed further below.

A search within the biological data is shown in Figure 13. Here the user has requested all records involving test system 3LGK5 (a breast cancer), and the compound whose NSC Number is 1895. Further, the tests must have been satisfactorily concluded - designated by a test status code (TSC) of 85. Only three entries satisfy these three criteria simultaneously; these are stored in a temporary file and then, upon command, incorporated into a tabular report, as shown.

MOLECULAR MANIPULATIONS BY COMPUTER

A wide variety of programs have been written to perform different operations upon chemical structures. Many of these are available to the DIS and are being used upon an experimental basis

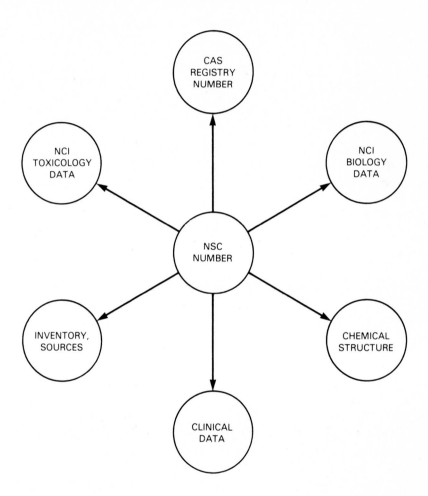

Fig. 9. Overview of the DIS, Central Role of the NSC Number.

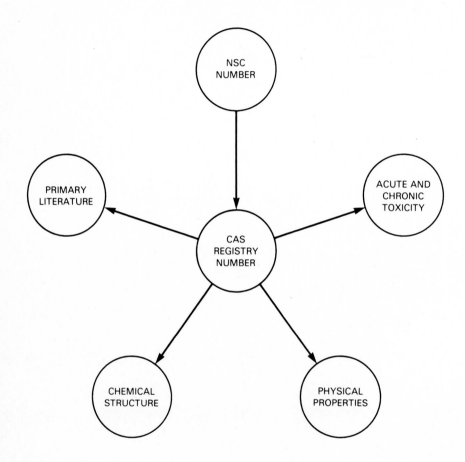

Fig. 10. Role of the CAS Registry Number in Information Retrieval.

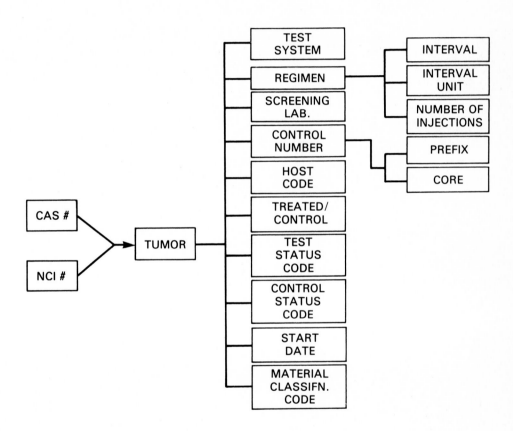

Fig. 11. Biological Data in the Drug Information System.

Option? <u>NPROBE</u>

Specify name (CR to exit):   <u>GUANAZOLE</u>

File 1,   1 compound having name: GUANAZOLE

      *** Caution: Name searched is ambiguous;
      *** desired compound may not have responded

Option? <u>SSHOW 1</u>

Entry   1   CAS RN 1455-77-2

Online:

    2 — CIS, EI Mass Spectrometry
    6 — CIS, Cambridge X-Ray Crystallography: DAMTRZ
  32 — NIOSH/CIS, RTECS: XZ4535000

Offline: 6 references $C_2H_5N_5$

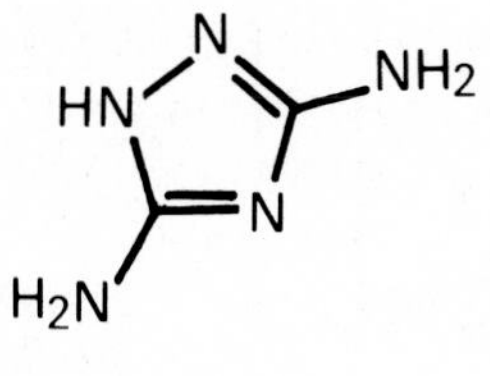

1H-1,2,4-Triazole-3,5-diamine (9CI)
Guanazole (VAN8CI)
NSC 1895
3,5-Diamino-s-triazole
3,5-Diamino-1,2,4-triazole
3,5-Diamino-1H-1,2,4-triazole

Retrieval by Name

Fig. 12. Search and Retrieval in the Drug Information System.

in an effort to support the various decisions that are made in the course of NCI drug development. Atomic coordinate data from x-ray diffraction experiments are available for over 30,000 organic molecules and access to these data can lead to displays such as those shown in Figures 14 and 15. In the first of these, the structure is retrieved from the CIS files with its CAS RN (550334). The DRAW command instructs the computer to locate and retrieve the atomic coordinate data for the compound and use them in developing the projection drawing of the molecule shown. The linear separation of various pairs of atoms can then be computed, upon request and these data can then be used to produce the space-filling model shown in Figure 15. Models of this sort, with the various distances, can be very useful in studies of biological activity.

The distribution of a compound between n-octanol and water, expressed as the log of the partition coefficient, is generally used as an indicator of the ease with which the molecule will be absorbed by a biological system. This parameter can be estimated as is shown in Figure 16. The command LFE/51172 instructs the computer to go to the record for that compound and check to see if atomic coordinate data are available for it. If so, the data are retrieved and used in the next step. If not, the program can take the two-dimensional structure and estimate the atomic coordinates for the most stable three-dimensional structure. The three-dimensional coordinates are then used by a program which estimates the linear free energy for the molecule in solution in n-octanol or in water. The log of the ratio of these free energies is the log P for the compound and this is reported to the user.

Finally, in an interesting conjunction of regulatory and research data, the inventory of chemicals in commerce that was complied by the EPA in connection with the Toxic Substances Control Act may be seen in Figure 17 to be useful to the DIS. In this example, the commercial status of benzimidazole (NSC # 759; CAS RN 51172) is retrieved and printed. This shows that the compound is manufactured, albeit in unknown volume, by Eastman Kodak in Rochester, NY. This company therefore may prove to be a useful source of the chemical.

GRAPHICS OUTPUT

As is clear from earlier sections of this paper, the DIS has a

major requirement for the automatic generation of reports of
different types from its various data areas. The chemical
structure of a molecule is frequently required as a part of these
reports and the system is therefore faced with the requirement
that it produce upon command, mixed graphics and textual output,
at high speed and with high quality.

Option? TSY/3LGK5 AND NSC/1185 AND TSC/85

File 1, Entries: 3

Option? TYPE 1/4/1-3

Freq	Int	Start Day	Total Days	Number of Injections	T/C	Test Status
Daily	1 Day	Day 1	10	10	84	85
Daily	1 Day	Day 1	10	10	106	85
Daily	1 Day	Day 1	10	10	109	85

Fig. 13. Search and Retrieval of Biological Data in the Drug
Information System.

A conventional impact printer cannot accomplish this and so
the DIS has been looking to the new generation of laser printers
for a solution to the problem. Laser printers write by passing a
modulated laser beam back and forth across a rotating
photoconductive drum. The beam is switched on and off by the
incoming data at a frequency which permits the writing of 180 dots
per inch in either the horizontal or the vertical directions. It
is therefore a pure graphics device. Drawings are fed to it,
rasterized and printed; alphanumeric characters are merely
special cases of graphics output. An added advantage of the laser

Option? DRAW 550334

File 1: Entry 1

NEBULR   CAS RN 550-33-4 Sub. 01

Isopurine, ribosyl
Nebularin(E)
Nebularine
NSC-65423
Purine ribonucleoside

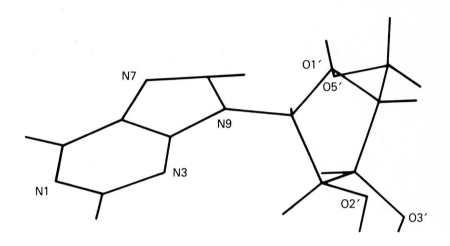

Option? DIST N1 O5´
　　N1 O5´ Distance: 7.38 Angstroms
Option? DIST N1 O3´
　　N1 O3´ Distance: 7.63 Angstroms
Option? DIST N1 O2´
　　N1 O2´ Distance: 6.51 Angstroms
Option? DIST N1 O1´
　　N1 O1´ Distance: 6.30 Angstroms

Computation of Geometric Parameters

Fig. 14. Use of Atomic Coordinates in Modelling.

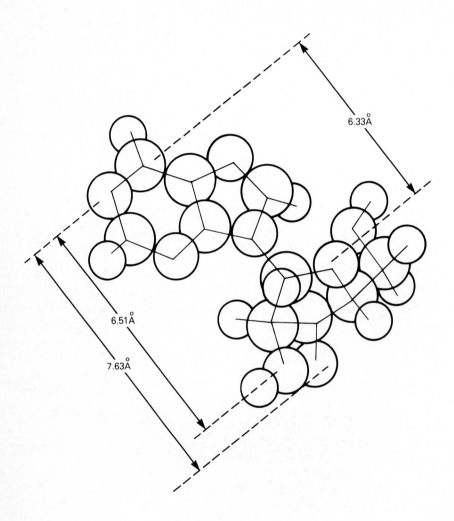

Fig. 15. Space-filling Model of Nebularine.

Option? LFE/51172

CAS RN 51-17-2   C7H6N2

```
              C  .  .  C
                  .               .
                      .               .
 N  *  *  C                    C
 %            .               .
 %                       .
 %           C  .  .  C
 %              *
 %          *
 C  %  %  N
```

1H-Benzimidazole (9CI)
Benzimidazole (8CI)
o-Benzimidazole
Azindole
Benziminazole

15 atoms

Sum(Polarizability):	5156.8	Sum(Polarizability)/N:	343.8
Sum(Charges)**2:	0.31	Sum(Charges)/N:	0.02

2 Rings Found: 6-Ring   5-Ring

F-H20:   − 10.91 Kcal/Mol   F-Oct: − 11.59 Kcal/Mol
Log Partition Coefficient: 0.50

Valence corrected Molecular Connectivity Indices:
XO: 4.834   X1:   2.848   X2:   0.928
X3 (PATH):   1.328   X3 (CLUSTER):   0.137   X3:   1.464

Molecular Weight: 118.1   Molecular Volume: 3164.7

Computation of Partition Coefficient

Fig. 16. Calculation of log P. Values.

Option? TPSHOW 1

```
                        imported
            manufactured.   :    .site limited
                  volume.  :  :  :   .size of plant
      ENTRY---CAS RN----- :  :  :  :  : --PLANT DATA ----------------
      1      51-17-2      N  Y  N  N  U    MID      910
```

                              PRD   − KODAK PARK DIVISION
                                         − ATTN: B. KLANDERMAN,
                                           BLDG. 26, 1669 LAKE AVENUE
                              CITY   − ROCHESTER
                              CNTY   − MONROE
                              STATE   − NY
                              ZIP   − 14650
                              EPAREG − 02

Manufacturing Data for NSC-1895 (Guanazole)

Fig. 17. Retrieval of Commercial Data on Benzimidazole.

printer is its speed. Most models print at a rate in excess of one page per second.

All the reports required from the DIS can be easily handled by a laser printer. The alphanumeric material poses no problem whatsoever; all characters are "drawn" with the help of macros and the character sets and fonts are limited only by the macros that are available to the printer. The chemical structures, it will be remembered, were generated and stored from the outset as vector sets. These can be readily rasterized by the processor within the printer and then printed along with all the other output data. An example of a report generated in this way is shown in Figure 18. The report begins with the full structure of a particularly complex chemical and then builds a table containing the requested data for that chemical. The output is generated by a single command, such as that shown in Figure 12; it is produced very quickly and is of a satisfactory quality for any purpose.

The power of a laser printer, operating as a graphics output device, facilitates a number of other tasks that are carried out routinely by the DIS. The printer is capable of delivering a logo, once it has access to the appropriate macro which describes the rasterized pattern. It can also print a signature using a digitized version of a previously entered template. With these two capabilities, the printer can accept a file of names and addresses and print a customized letter, on official letterhead and already signed, to every individual on the list. For an organization which must generate, sign and mail several hundred form letters per month, this has obvious possibilities.

In another potential application of the laser printer, some study is continuing of the material handling procedures within the DIS. When the program receives a new chemical, it is assigned an NSC Number and, amongst other things, a series of gummed labels bearing that number is printed and used to label the samples. A laser printer can print labels bearing the Universal Product Code (UPC) that is used in merchandising, and the possibility that NCI samples might be physically stored and labelled with such a machine-readable identifier is intriguing. There is in fact, no reason why all samples could not be stored in this way and material inventories checked with an electronic balance which is online to a central computer. The balance measures and transmits the weight of sample in a tared container and a light pen can be

NSC 3053

NSC 003053   QUERY NUMBER 204

TEST SYSTEM SCHEDULE	RTE	BEST					REPRODUCIBLE					ACTIVE RESP	DOSE RESP
		EVAL DAY	T/C	DOSE	VEH/TOT	CURES/TOT	EVAL DAY	T/C	DOSE	VEH	CURES/TOT		
OTHER	IP	2	148	0.03	9	0/06						1	-
OTHER	IP	046	158	203.00M	2	0/10	046	144	104.00M	2	0/10	2	-
OTHER	IP	046	136	1.88	7	0/06	046	132	6.00	8	0/06	2	-
DAILY, TWICE A DAY	IP	2	142	12.50	2	0/06		130	27.00M	8	0/06	10	-
Q1D DAILY TO DEATH	IP	2	153	50.00M	7	0/10		151	150.00M	7	0/06	7	14
Q1D DAY 1 ONLY	IP	030	140	256.00M	2	0/10	030	144	0.05	7	0/06	2	-
Q1D DAYS 1 - 7	IP	030	162	0.10	2	0/06	068	168	64.00M	9	0/08	9	-
Q1D DAYS 1 - 9	IP	068	170	100.00M	2	0/08	2	143	50.00M	2	0/10	19	27
Q1D DAYS 1 -10	IP	2	167	50.00M	2	0/06	060	141	0.50	2	0/06	3	-
Q1D DAY 2 ONLY	IP	060	181	0.50	2	1/10	046	145	0.90	7	0/10	2	-
Q1D DAYS 2 -16	IP	046	153	67.00M	7	0/10	046	176	100.00M	2	0/10	2	-
Q1D DAYS 5 UNTIL DEATH	IP	2	114	0.03	7	0/06		111	0.20	7	0/06	1	-
Q1D DAY 5 ONLY	IP	2	122	175.00M	7	0/06						2	-
Q2D DAYS 2 THRU 16	IP	046	150	150.00M	7	0/10		144	37.00M	2	0/06	1	14
Q03D UNTIL DEATH	IP	2	161	300.00M	7	0/06						2	-
Q3D DAYS 1,4&7	IP	030	165	0.10	2	0/06	030	164	0.20	7	0/06	1	-
Q4D DAYS 1,5&9	IP	030	170	75.00M	2	1/08						10	12

Fig. 18. DIS Report Printed on a Laser Printer.

used to transmit the corresponding NSC Number. In this way, an
accurate inventory of material could be established and
maintained.

CONCLUSION

The NCI Drug Information System uses a very large and
continually changing database. Further, there are a number of
requirements of the DIS which are such that some special effort is
necessary to produce a satisfactory computer system.

Efforts to address these problems lead to explorations in
available technology and as such, may be of interest to the
information community in general. This paper has tried to describe
some of these areas; working systems and planned systems have been
freely mingled in the narrative; the techniques described all
exist, but their application to the DIS is not in all cases
complete. Nevertheless, these are the directions which are being
taken by this large and dynamic system in an attempt to address
the continually more searching requests that are made of it by
users.

S.R. Heller and R. Potenzone, Jr. (Editors), *Computer Applications in Chemistry*
© 1983 Elsevier Science Publishers B.V., Amsterdam — Printed in The Netherlands

A CENTRALIZED COMPUTER SYSTEM WITH DISTRIBUTED SATELLITES FOR THE ACQUISITION AND INTERPRETATION OF ANALYTICAL DATA

ENGELBERT ZIEGLER

Max-Planck-Institut fur Kohlenforschung, Mulheim/Ruhr, GERMANY

SUMMARY

For more than 12 years a central DEC10- computer system had served for all data processing requirements in chemical research at the Max-Planck-Institutes in Mulheim (Germany), including real-time data acquisition from a multitude of analytical instruments. In 1981 it was replaced by a hierarchically structured computer system: The real-time tasks are performed by local mini computer sattellites that are tailored to special requirements in the laboratories. All higher-level data processing and data storage demands are covered by the central system, consisting of two VAX11/780- computers, to which all satellites as well as all user terminals are connected. Some of the software packages developed for this system are described.

Background

Over the years many different approaches for laboratory computerization have been published. More than 15 years ago, when computer hardware was very expensive, multi-instrument multi-method systems had been implemented where one computer serviced a variety of instruments of different kind. Those early systems suffered from a lack of sufficient hardware resources; moreover the operating system software available for these computers was not especially suited to allow the time-shared use of the computer for many independently operating analytical methods.

When lower cost mini computers became available, "dedicated" systems could be built where one computer serviced one analytical instrument or several instruments of one kind, e.g. a group of chromatographs. These dedicated mini computer systems could be

tailored to the specific requirements of an instrument and were
able to perform extensive instrumental control functions besides
the   task   of signal sampling. For several   years, however,
supporting system software for mini computers was rather sparse
and difficult to use for an analytical chemist.

Another major drawback was caused by the limitations in
processing power, especially for more complex, large programs, and
in available storage capacity for collecting and archiving of
analytical data.

An alternate way to computerize a multitude of analytical
laboratories had been chosen at the Max-Planck-Institutes in
Mulheim/Ruhr. (The Max-Planck-Institut fur Kohlenforschung and the
MPI fur Strahlenchemie are located on the same area. Research
activities are focused on organometallic chemistry and radiation
chemistry respectively. The two institutes share analytical
laboratories, like X-ray diffractometry, chromatography and mass
spectrometry, and also maintain a common computing center and data
processing department.) In 1968 a centralized DECsystem-10
computer system had been brought into operation. Its major
advantage over dedicated systems was caused by its powerful
time-sharing operating system with extensive features to
facilitate interactive dialog programming, by its processing speed
and its data storage peripherals. It was used for all data
processing requirements in a chemical research environment, for
'batch'-type computational tasks as well as for dialog- oriented
tasks, with more than 50 terminals connected in its final
configuration. Besides this time-shared usage, however, the system
was designed to take care of the real-time data acquisition
requirements: more than 60 'slow' instruments with data rates of
up   to   20 A/D-samples per   instrument   per second, mostly
chromatographs, were directly connected over distances between 60m
and 200m to a central high-quality multiplexed A/D-converter with
auto-gain-ranging capability. Furthermore up   to 4 'fast'
instruments with data rates between 1000 and 10000 samples per
second, especially fast-scan, low-resolution mass spectrometers
transferred data via a second A/D-system to the central computer.
The system had been described in several publications and became
known as 'Mulheim Computer System'. (ref.1)

Even at the time when this approach had been chosen it was
already obvious that a hierarchically structured computer system

would have been the preferred solution: the real time tasks would be performed by mini computer satellites, whereas the more demanding data processing tasks and the data storage requirements would be covered by a large central machine to which all satellites were connected. Such a hierarchical system, however, would have been by far too expensive, and useful networking software was not available at that time. Today, more than 12 years later, the rapid progress in computer technology resulted in the integration of all measuring and all instrumental control functions into the instrument by means of dedicated micro processors. Higher level data processing, however, as well as the need for centralized data banks still require some central facility with sufficient computing power and capacity for data storage. Furthermore, older instruments without integrated microprocessor logic are still in existence and are too expensive to be replaced. For these reasons a hierarchical system is still an ideal solution for the computerization of analytical instruments.

In 1981 the old Mulheim System was finally replaced by a new one. In the following section the essential design criteria of this successor system that are based on the experiences with the old system are discussed.

Experiences with the old computer system: consequences for the new system

Over time the old system had been upgraded and stepwise expanded to the following end configuration:

KI10- processor with 288 K 36-bit words core memory

600 MB of disk storage

2 magnetic tape drives and 4 DECtape- drives, line printer, plotter, card reader

A/D- system for 64 'slow' channels with auto-gain ranging (1 uV to 10 V)

A/D- system for 4 'fast' channels (40 uV to 10V)

56 asynchroneous serial ports for terminals, terminal-like devices and mini- computer connections

For some other instruments, like modern NMR-spectrometers or X-ray diffractometers with built-in stand-alone computer capabilities no analog-signal connections were necessary; file transfer protocols via serial lines had been established instead. Furthermore a few mini computers used in radiation chemistry experiments were connected to the central machine.

In the last years of operation about 5000 to 5500 hours CPU-time had been consumed annually, mostly for quantum chemistry work (ca. 40%) and X-ray structure determinations (ca. 30%), whereas ca. 13% had taken up by the analytical and the spectroscopy laboratories. System availability during the main working hours averaged to 98.5% with ca. 3 unscheduled system restarts per month. The load on the real-time part is illustrated by ca. 40,000 chromatograms and ca. 400,000 mass spectra (in repetitive scan mode) processed each year.

Reliability and availability of this system was insufficient only during the first two years of operation. With improving vendor maintenance and especially after adding more hardware computer availability was no longer a problem: with a total of 4 or 6 disk drives, for instance, the failure of one drive did no longer shut down the entire system. The frequency of hardware failures of a large system is, of course, higher than with a small mini computer system. On the other side higher redundancy of the hardware, its operation by knowledgeable personnel supported by sophisticated systems software lead to shorter down times: in many cases the system can be made operational within a matter of minutes.

There were other features that proved to be of importance to the users:

a) The operating system allowed for general-purpose time-sharing; via dialog terminals many simultaneous users were able to edit, to test, and to execute programs interactively. There were no stringent restrictions in program sizes nor in available computing power. It was possible to structure the software for the real-time application as to simulate multiple independent dedicated computer systems. Therefore the necessity to use dedicated stand-alone systems for real-time applications with analytical instruments had been minimized.

b) The very specialized computer related knowhow was concentrated within a small team which was responsible for operating the system, for developing software tools needed by special applications and for assisting users. Therefore the latter were not forced to learn too many computer specific details and could concentrate on their applications.

c) Terminals were easily accessible for all interested persons. For this reason up to now no 'personal computers' have been purchased by any departments. The terminal to the central computer was easier to operate and offered by far more computing capacity and flexibility than any of today's personal computers.

d) Within the centralized system the users had access to high-performance, high-capacity peripherals that are normally not affordable for distributed small systems.

e) For the real-time data acquisition expensive high-performance A/D-converters with multiplexers and auto-gain ranging amplifiers could be applied, resulting in high dynamic range of input voltages to be sampled. High dynamic range was considered an essential requirement for our analytical instrumentation: the connection to a computer must not deteriorate the performance of an instrument.

Besides this list of positive characteristics of the old 'Mulheim System' to be asked also from a possible successor system, there were also several aspects where improvements could be achieved:

a) The central computer should be released from all direct interactions with analytical instruments. Closed loop instrument controls are difficult to incorporate into a large centralized system, - especially, if the algorithms used in closed loop actions should change too often.

b) The central computer should be operated with standard operating systems software as supplied by the computer manufacturer.
In the old system any changes to the operating system had to be carried through all versions of the software of the manufacturer.

c) The analytical instruments to be connected to the computer

are located in different buildings of both institutes. In
the old 'Mulheim System' the analog-signals from the
instruments were transferred through shielded cables and
over long distances to the A/D-hardware in the computing
center. Since the new computer system was intended to be
installed within a new building a considerable amount of
recabling would have been required by this old approach.

d) If too many instruments, especially simultaneous
fast-scanning ones, are to be serviced by a central system,
the other users of the computing center would experience
performance decreases.

For this reason (and because the transition to a new system
could be foreseen) one of the mass spectrometers had been
connected to a PDP11/34 minicomputer already several years ago.

All these desired improvements resulted in the design of the
distributed system as described in the following paragraphs: the
real-time servicing of instruments is off-loaded to small
sattelite computers located decentralized in the main analytical
laboratories. All the user-oriented software for the processing of
real-time data, however, is operated on the central machine with
all user terminals connected to it. In this way the desired
improvements could be achieved without losing the advantages of
the previous computer system.

Hardware configuration of the new central system
The new computer system has been designed around two VAX11/780
CPU's (manufacturer: Digital Equipment Corp.) (Figure 1). The two
CPU's are supplied with 3 and 3.5 MBytes of main memory,
respectively; total disk capacity is 1900 Mbytes, consisting of 5
SMD-drives with 256 MB capacity each, 1 FMD disk with 512 MB,
connected to the SBI-bus, and 4 cassette drives with 28 MB each,
connected to the UNIBUS. The only one magnetic tape unit (1600
bpi, 125 ips) is used for data exchange with other installations
and for the periodic backup-procedures of the disk storage. The
two systems execute separate copies of the VMS operating system,
but are loosely coupled via a 1 MBd serial (DECnet-) link and
share the main disk drives: each system may read all of the 6
drives that are connected to the SBI- bus, but has access for
writing to only 3 of them.

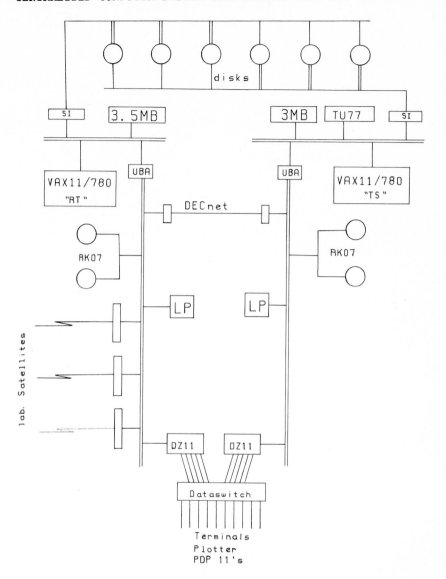

Fig. 1. Hardware configuration of the central computer system in Mulheim with two VAX11/780-CPU's.

About 90 terminals are connected via asynchroneous serial lines to a central communication control unit ('DATASWITCH', Infotron Corp.). This 'intelligent' control unit allows the user when logging into the system to choose within classes of computer ports and especially to select one of the two CPU's for his dialog

session. Port classes are defined transparently: most users do not
know with which one of the two systems they are working; if, for
any reasons, one of the systems becomes unavailable, easy and
quick reconfiguration of the remaining system is possible.

   The serial lines are also used to connect decentralized
devices like plotters and digitizer tablets as well as 6 PDP11
mini computers (3 RSX11M- , 1 RT11- system, and 2 machines without
a standard operating system that are down-line loaded from the
central system) (Figure 2). These mini computers are used for
special applications, like collecting data from X-ray
diffractometers, from mass spectrometers, or for monitoring and
evaluating pulsed radiolysis measurements. The accumulated data
files are transferred to the central system for further processing
and for archiving. The necessary file transfer software has been
written tailored to these special requirements, avoiding unneeded
features and the high internal overhead and costs of commercially
available network products

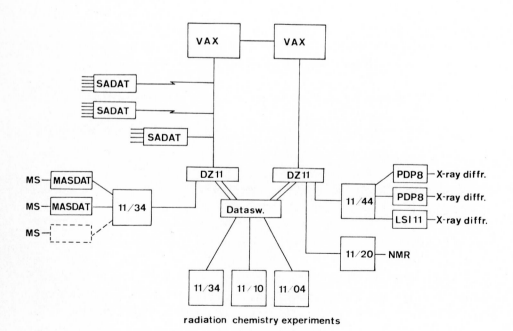

radiation chemistry experiments

Fig. 2. Laboratory satellites connected to the central system.

Other decentralized satellite computers, for instance for chromatography and for mass spectrometry, are connected through coaxial cables to DMA-interfaces of the central system, as described in the next session.

Tightly coupled real-time satellites

As already mentioned one of the objectives of the new system was to free the central computer from real-time tasks. This is accomplished by small satellite computers, which take care of ADC-sampling, start/stop- requests of instruments and perform all necessary instrument control and timing functions. All collected instrumental data are first buffered in the main memory of the satellite and then transferred blockwise to the central VAX-system via a coaxial cable and a DMA- interface. Within the main memory of the VAX a permanently resident section is defined to hold a ring of data buffers and parameter tables. The active 'master' part is assigned to the satellite: data blocks and parameter tables are read and written from and into VAX- memory without the VAX knowing about it. Only on the completion of a data block transfer an interrupt to the VAX- cpu is generated in order to trigger some action like writing a block of data to disk memory. Actions of the satellite are either initiated by external signals (e.g. start/stop) from instruments or by commands issued by VAX- programs, for instance if new data acquisition parameters have been specified by a user through a dialog via a VAX-terminal. By this design the overhead at the VAX- side is minimized. In fact, the acquisition and transfer to disk of a block of real-time data require less VAX- resources than copying a block of data from one disk file to another one.

The satellites are based on LSI11- computers (manufacturer: Digital Equipment). They are programmed in assembler language according to their application without an underlying operating system. When a satellite is powered up, or after not reacting to VAX- commands its program is loaded via the coaxial connection through a DMA- transfer initiated by the VAX- side. (When the system went into operation the DMA- interface was not yet available; therefore a temporary solution had been constructed where the satellites were connected to the host via serial lines.)

Currently two different implementations of real-time satellites are in existence:

(a) the SADAT- system ("SAtellite for Data Acquisition and Transfer"), which is mainly used for multiple 'slow' instruments, like chromatographs and some optical spectrometers;

(b) the MASDAT- system ("MAss Spectrometry Data Acquisition and Transfer"), where a satellite is dedicated to one low resolution fast-scan mass spectrometer.

These two systems will be shortly described in the following paragraphs. Other types of satellites are still in the design phase or under discussion.

Connection of 'slow' analytical instruments. (SADAT- system.)

For the majority of analytical instruments digitization rates of 25 data points per second or less are sufficient. Multiple instruments of this type, e.g. chromatographs or most optical spectrometers, may be connected to a single satellite computer. In its present hardware implementation a SADAT- system allows for up to 48 input channels with analog signals. The data rate r may be selected for each channel individually, with

$$r = f / 2^n \ , \ (n = 1, 2, \ldots, 16),$$

where f is the power line frequency (50 or 60 Hz). Tight synchronization of data sampling with the line frequency is necessary in order to avoid signal distortions caused by ripple.

Time-linear A/D- sampling certainly is the most important task of the SADAT- satellite. Many other functions, however, have been designed into the system. The following is a list of the existent features:

1) Sampling of (up to 48) analog inputs
   a) with fixed preselected data rates;
   b) triggered by external stepping signals;
2) digital sampling of instruments with RS232- (serial-) interface;
3) digital sampling of instruments with IEEE-bus interface;
4) continuous time-linear sampling ('monitoring') of signals (e.g. room temperature) without explicit start-/stop- requests;
5) servicing of start-/stop- requests from push- buttons;
6) 'stepping' of instruments
   a) by outputting a specified number of digital pulses

requested by a program in the central computer;
b) by outputting a single pulse after each sampling of the
   input channel;
7) time-dependent instrument control:
   a) periodical switching of relays;
   b) switching of a relay at a prespecified time after start
      of an instrument;
8) closed-loop instrument control (via relay switching) by
   software routines that are loadable into the satellite;
9) immediate starts and stops of instruments, or switching of
   relays as requested by programs in the central computer;
10) setup of instrument parameters via RS232 instrument
    interfaces initiated by the central computer;
11) setup via IEEE- instrument interfaces;
12) backup with intermediate buffer storage on external media
    (cassette tape), if the central computer is temporarily
    inaccessible.

Besides servicing instruments the DMA- connection of the
satellite to the central computer provides a convenient link to
implement fast file transfers between some minicomputer systems
and the central computer: the SADAT- satellite acts as an
interface with the minicomputer connected to the satellite by a
serial RS232- line. The advantage of this solution is to free the
central computer from too much character handling overhead
associated normally with serial- line file transfers. It should be
mentioned that the current hardware implementation (Figure 3) does
not include yet software functions 3 and 11. The hardware consists
of
    LSI11/2 with KEV11 extended arithmetic,
    64 KB of main memory,
    2 DRV11J for 128 bits of digital I/O,
    1 DLV11J with 4 RS232- interfaces,
    1 GMS11 DMA-interface to connect the satellite to the
      VAX-Unibus over coaxial cables (manufacturer- GDV,Tubingen),
    RTP7480 wide range analog input system (13 bits precision, 13
      gain ranges, 5 uV noise, ca. 160 samples/s throughput),
      or optionally:
    TP7486 high-speed wide range analog input system (10 uV
      noise, 3300 sample/s).

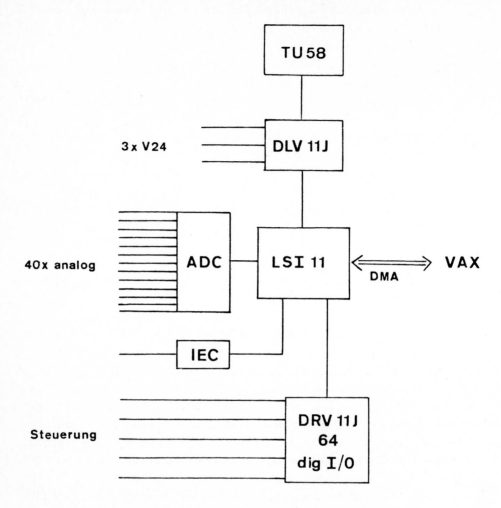

Fig. 3. Hardware implementation of a "SADAT"- laboratory satellite
for servicing of slow instruments.

   The ADC- systems (manufacturer: Computer Products, Fort
Lauderdale) are supplied with auto-gain ranging hardware and allow
for a dynamic voltage range of ca. 10^6 or better (signals between
10 uV and 10 V).
   Currently three SADAT- systems are implemented: two of them
are mainly used for chromatography (one with 42 chromatographs and
1 slow scanning mass spectrometer used for quantitative analysis
of gaseous mixtures; one for 25 chromatographs and several
photometers), a third one is used with optical spectrometers, 2

chromatographs and (not yet operational) 1 ultra-centrifugue.

Connection of fast-scan mass spectrometers (MASDAT- satellite)

Fast-scan low-resolution mass spectrometers are typically used in GC/MS combinations or for temperature controlled fractionated evaporation of solids. In both cases the spectrometer will be operated in repetitive-scan mode, e.g. scanning a spectrum of 2 s duration every 3 s with a data rate of typically 5000 or 10000 samples/s.

In the original 'Mulheim System' the mass spectrometers had been connected directly to the central DECsystem-10 computer. In an effort to off-load some of the real-time applications from the central machine a mass spectrometry data system had been built based on a PDP11/34 minicomputer system (with RSX11M operating software) with a LSI11 satellite computer used for the acquisition and reduction of spectra. This data system, also supplied with spectra evaluation software, became operational for one mass spectrometer in 1978. After installation of the VAX- system and the removal of the DECsystem-10 two spectrometers, each with its own LSI11- satellite, were serviced by this PDP11 system. At the same time an improved version of the evaluation software including extensive library search facilities had been installed on the VAX-system, and therefore the PDP11/34 was used exclusively for measuring spectra. After completion of a measurement a series of spectra is transferred over a serial-line connection to the VAX-system for further processing. (It is intended, however, to connect the MASDAT- satellites in the near future directly to the VAX- Unibus, thus making the PDP11/34 obsolete.)

The MASDAT- satellite performs the following functions for its associated spectrometer:

(1) setting of ADC parameters;
(2) switching between alternative ionization voltages;
(3) start and stop of spectrometer scan;
(4) acquisition of mass spectra;
(5) monitoring of up to 3 additional signals like temperature, pressure, total ionization
(6) reduction of 'raw data' to peak candidates;
(7) transfer of reduced and/or unreduced spectra to the host computer.

The satellite operates in one of three different modes: transferring unreduced spectra, transferring reduced spectra, or transferring unreduced as well as reduced data for a spectrum.

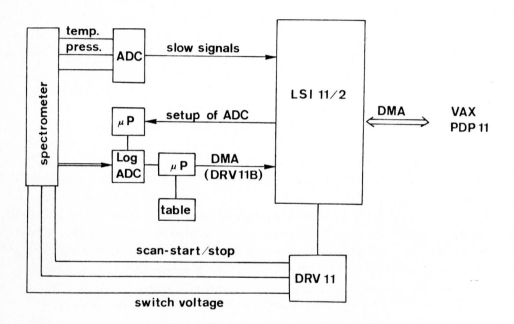

Fig. 4. Hardware implementation of a MSDAT- satellite for fast-scan low-resolution mass spectrometry.

The hardware implementation (Figure 4) of a MASDAT- satellite consists of

LSI11/2 processor with KEV11 arithmetic,

64 KBytes main memory,

1 DRV11 for 16 bits digital I/O,

DT2764 ADC, 16 bits resolution, for the 3 slow channels,

LLG- spectrometer- interface (especially manufactured by Busch Elektronik, Waldbronn) based on a microprocessor controlled logerithmic ADC (Analogic),

1 DRV11B for DMA- transfer from spectrometer interface into

LSI11,

GMS11 for coaxial DMA- connection of the satellite to the central computer.

The spectrometer interface is capable of sampling the signal with extremely low noise within a signal range of 10 uV to 8V, with the original logarithmic ADC- result being translated into a linear scale by the built-in microprocessor. Sampling rates of 1.25, 2.5, 5, 6.67, 10 and 20 KHz are possible.

Currently two MASDAT- satellites are in operation. (More details will be published elsewhere.)

Software overview

For several application areas software from external sources is being used. In many of these cases, however, extensive modifications and expansions have been made, for instance in the field of x-ray structure determinations and electron density investigations, or for quantum chemistry calculations. In other areas, especially in mass spectrometry, chromatography and optical spectroscopy, no external software at all is in use. In the new computer system almost all programs are written in FORTRAN-77 language. Because of the large virtual address space provided by the VAX- architecture no restrictions in program sizes had to be observed. Both facts contribute to faster program development and improved designs of user dialogs, e.g. in the areas of free-format input, error messages, and "HELP"- facilities.

In the following section the components of the chromatography and of the mass spectrometry software packages are shortly described. Both packages can be transported to other VAX- systems, for instance to very small ones dedicated to the respective application.

Chromatography software

As described already the SADAT satellites will take care of the data acquisition part. All data evaluation and all user dialogs, however, are performed by software within the central VAX- system.

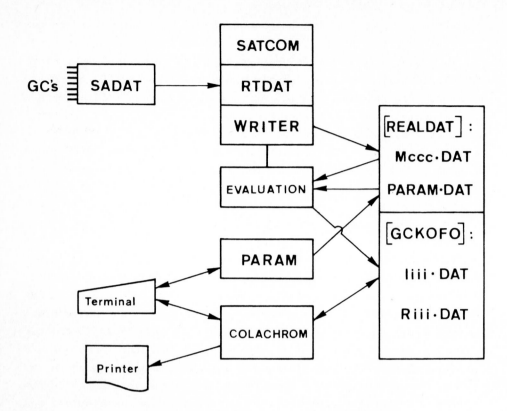

Fig. 5. Structure of chromatography software package.

The different software components are (Figure 5)

(1)   Parameter specification (program 'PARAM'): maximum runtime,
      data rate, instrument parameters, sample identification, etc.,
      may be entered through a programmed dialog or via a set of
      special commands.

(2)   Satellite   communication ('SATCOM')   and   data transfer
      ('WRITER') of blocks of unreduced ADC- data that were written
      by the satellite via DMA directly into a resident section
      ('RTDAT') within VAX- memory.

(3)   Automatic   evaluation   of the unreduced   data after   a
      chromatogram has been stopped (either by push- button or on
      elapsed runtime); a report consisting of an extended peak list
      is written to the disk and optionally to a local laboratory
      report printer.

(4)   A refined evaluation of the report is performed in dialog

mode. The COmmand LAnguage for CHROMatography, COLACHROM, has been developed to allow the definition of standards for quantitative analysis as well as for retention times in form of Kovats indices or relative retention times, etc., to select or to suppress peaks for the final report, to define thresholds for peak area percentages or weights, to determine column parameters, like plate numbers, coating efficiency, etc.. Another set of commands supports the graphical display of unreduced chromatograms, for instance with the computer-calculated baseline drawn in, and the use of the graphical cursor, e.g. in order to correct the baseline findings or to visually identify standards, etc.. Other commands allow to repeat the peak analysis with a changed set of peak search parameters or to repeat part of the peak search only, for instance if a user-corrected baseline is to be used. In order to facilitate the building of tables of chromatographic data and the comparison of peak lists with such tables various further commands are available. In total more than 80 commands exist for processing report data or unreduced ADC- data in many variations. Some of these commands are used to print the final report at a user terminal, at a laboratory printer or at the computing center printer, or to archive the report on disk storage.

(5) A bookkeeping and accounting system is embedded into the software package for generating analysis statistics and for keeping track of computer usage.

Mass spectrometry software

Measurement of spectra (i.e. series of spectra) is performed through the MASDAT- satellite as described above. During the measurement an interscan- report is produced to inform the operator of the spectrometer on the progress of the experiment.

The completed series if spectra are further processed:
(1) normally an automatic extraction and correction of 'relevant' spectra out of a series of spectra will be initiated (ref. 2);
(2) the extracted ones will be compared to one or more libraries of spectra (SISCOM- software, ref. 3) resulting in 'hit-lists' for all relevant spectra;
(3) instead of the automatic procedure all steps in (1) and (2)

may be executed interactively with the analyst specifying
through the programmed dialog exactly the operations to be
applied to selected spectra;
(4) part of the software package are programs for the computer
input of chemical structures and their pictorial reproduction
(ref. 4), for instance in connection with a hit- list of the
library search,
(5) utility programs, for instance to calculate the isotope
distribution of fragments, are available in order to work with
unreduced spectra (e.g. for diagnostic purposes), to produce a
variety of graphical representations and to allow statistical
investigations.

CONCLUSIONS

Hierarchically structured systems like the one described in
this paper are ideally suited for laboratory applications: local
mini- and micro-computer satellites can be tailored according to
special measurement and control requirements of a certain
instrument or a group of instruments. All higher-level data
processing as well as all data storage tasks are performed by the
central system, with all user terminals connected to it. Thus all
more demanding software development is supported by the power and
the resources of a modern computer system - leading to better and
more intelligent programs with improved "human interface".

ACKNOWLEDGEMENTS

Involved in the work described in this paper have been my
coworkers K. Boll, H. Lenk, H.D. Schmitz, Dr. B. Weimann and I.
Wronka. Many helpful discussions with the users of the computer
system, e.g. Dr. D. Henneberg, Dr. G.Schomburg, U. Hausig, F.
Weeke, contributed to the software developed.

REFERENCES

1 E. Ziegler, D. Henneberg, G. Schomburg: Analyt. Chem. 42,
(1970), 51A; E. Ziegler, D. Henneberg, G. Schomburg, B. Weimann:
several articles in angew, Chemie intern, Edit. 11, 348 ff.
2 D. Henneberg, H. Damen, B. Weimann: Adv Mass Spec. 7 (1978) 975.
3 H. Damen, D. Henneberg, B. Weimann: Analyt. Chim. Acta 103
(1978) 289.
4 E. Ziegler, K. Boll: Analyt. Chim. Acta 103 (1978) 237.

S.R. Heller and R. Potenzone, Jr. (Editors), *Computer Applications in Chemistry*
© 1983 Elsevier Science Publishers B.V., Amsterdam — Printed in The Netherlands

TEACHING WITH A MICROCOMPUTER

STANLEY SMITH
University of Illinois, Urbana, Illinois 61801

Experience with thousands of students over the past decade clearly indicates that computer-assisted instruction (CAI) works. Students using appropriate CAI programs as a required part of courses do better on written examinations (ref. 1) and make fewer errors in laboratory work (ref. 2) than those who have only traditional instruction. In addition students like using the computer as a learning tool (ref. 3).

It is now possible (ref. 4) to deliver high quality computer-assisted instruction on systems ranging from a home computer with courseware on a diskette to a terminal which is on-line to a large computer system such as PLATO (ref. 5) which serves over a thousand graphic terminals in a nation wide network (ref. 6).

The basic requirements of a computer system used for CAI are the ability to deliver the required lesson to the student, to put text on the screen along with complex drawings, figures, graphs and formulas, and to judge student input to questions and make the appropriate response. This is illustrated in Figure 1, which is part of a course in general chemistry (ref. 7).

These displays illustrate the resolution which is possible with the Apple 280 by 192 bit mapped display.

In this case, the lesson was selected by the student by putting a diskette into the disk drive of an Apple (ref. 8) microcomputer and turning on the computer. In this lesson the student is to explore the relationship between pressure and volume of a gas at constant temperature. The student can move the piston in and out and record volumes from the scale on the piston pressures from the manometer. Results of a typical experiment are shown in Figure 2 as a plot of pressure vs. volume. In this way, the student discovers the relationship between pressure and volume

of a gas through simulated experiments. The complete set of
lessons includes experiments on volume as a function of
temperature at constant pressure and pressure as a function of
temperature at constant volume. The experimental simulations are
reinforced by having students work typical gas law problems.

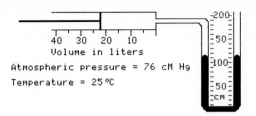

Fig. 1. One display from a lesson on ideal gases by Smith and
Chabay which runs on an Apple II microcomputer (ref. 7).

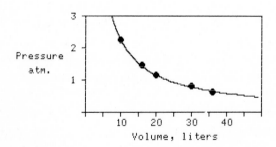

Fig. 2. Plot of pressure vs. volume in a simulated experiment from
a lesson by Smith and Chabay (ref. 7).

The display in Figure 3, which is from a lesson on multistep
aromatic synthesis (ref. 9), serves to illustrate the ability of a
personal computer to draw structural formulas and to determine and
draw the structure of the product of these chemical reactions. In
this program students suggest ways to prepare specific compounds
by selecting a starting material and then selecting one of twenty
reagents. The microcomputer determines the structure of the
product of the reaction and draws it on the screen. Because of the
speed of the microprocessor, response is nearly instantaneous.

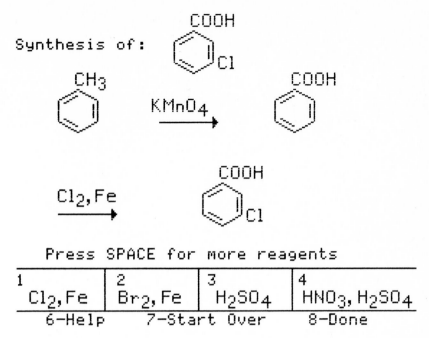

Fig. 3. A complete synthesis of m-chlorobenzoic acid. The student started with toluene and elected to first oxidize and then chlorinate. The program determines the product of each reaction so alternative approaches to the synthesis are accepted (ref. 9).

An introductory course in polymer chemistry has been written (ref. 11) which runs on either an Apple microcomputer or on the PLATO(5) system. This allows some direct comparisons of between big computer systems dedicated to delivering instructional material and a small personal computer.

A typical display from the microcomputer version of the polymer chemistry course is shown in Figure 4 while the on-line PLATO version is shown in Figure 5. The difference in resolution of the Apple (280 by 192) vs. PLATO (512 by 512) can be seen by comparison of these two figures. The PLATO program written in TUTOR, which is a compiled language, requires about 5 seconds (the time depends on the number of users of the system) to compute and plot one of these curves while the Apple program, running in interpreted BASIC, takes about 20 seconds. The major cause of the time difference in the two systems is the relative speed with which they do floating point arithmetic.

Weight Fraction Distribution of Polymer Chains.

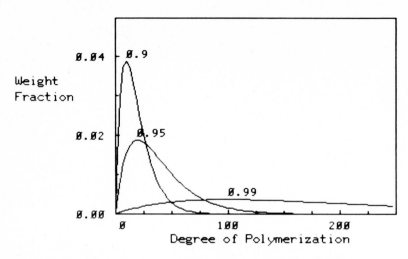

Figure   4. Display from   a   course in polymer chemistry   (ref. 11)
            running on   an   Apple   microcomputer. This system   has a
            resolution of   280 by   192 dots.   A similar display from
            the PLATO system is shown in Figure 5.

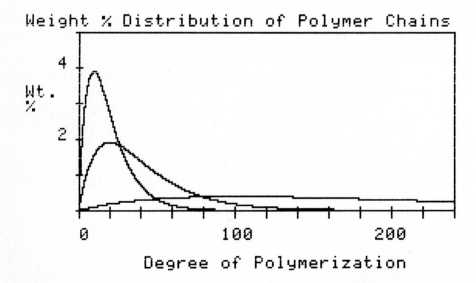

Figure   5.   Display   from a course in polymer chemistry (ref.   11)
            running   on   the   PLATO system.   The PLATO system has   a
            resolution of 512 by 512 dots. A similar   display on   an
            Apple microcomputer is shown in Figure 4.

It is very important for a CAI system to respond in a helpful way to student errors. Hence the ability to automatically detect and point out minor misspellings was a major contribution of the PLATO system (ref. 11) to the development of CAI. This is illustrated in Figure 6 taken from the PLATO version of the polymer course (ref. 10). Here the word poly(methyl methacrylate) is underlined to indicate that it is not just wrong but rather misspelled.

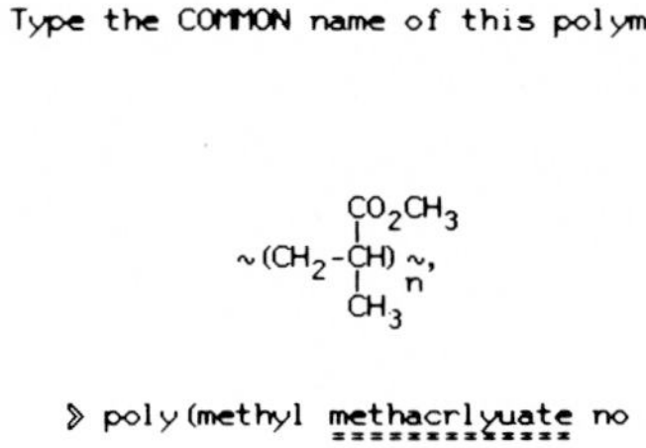

Type the COMMON name of this polymer:

$$CO_2CH_3$$
$$\sim(CH_2-CH)\sim,$$
$$CH_3 \quad n$$

> poly(methyl methacrlyuate no

Figure 6. Part of a lesson on polymer chemistry (ref. 10) running
on the on-line PLATO (ref. 5) system. The system
automatically underlines words which are misspelled
(ref. 11).

The same question, running on an Apple II microcomputer is shown in Figure 7. Here, in addition to detecting that the word is misspelled, the Apple program is able to give specific feedback on the nature of the error right down to the nearest letter! In this case the A is underlined to indicate that it isn't the right letter (should be E), the markup under the LY shows that the letters are inverted. There is an X under the U because it doesn't belong there and the arrow at the end indicates something, namely), is missing at the end.

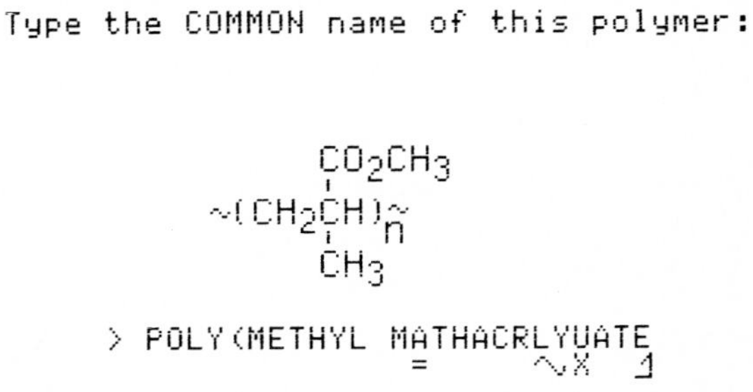

Type the COMMON name of this polymer:

$$CO_2CH_3$$
$$\sim(CH_2CH)\sim$$
$$CH_3 \quad n$$

> POLY(METHYL MATHACRLYUATE

Figure 7. Example (ref. 10) of the detailed spelling markup now
available on an Apple II computer (ref. 12).

This ability to provide detailed help to users on spelling and typing errors in BASIC programs running on an Apple II computer was developed by Paul Tenczar (ref. 12) and represents a major advance in computer-assisted instruction.

The spelling judger illustrated in Figure 8 has been implemented on an Apple by adding new commands to the standard BASIC language. These commands have been selected to provide those features most needed in the development of high quality computer-based teaching programs.

Here is a BASIC program which illustrates the use of a few of these added commands.

```
100 &N : REM erase high resolution screen
110 VTAB 5: HTAB 10
120 PRINT "Type the name of NaCl"
130 &I : REM student input
140 &A"sodium chloride" : REM correct answer
150 &M : REM do spelling markup
160 IF ZW% THEN 130 :REM if wrong try again
```

This simple program provides the spelling analysis illustrated in Figure 7. The command &N on line 100 erases the screen, attaches the high resolution character plotter and initializes the input command, &I. The answer command, &A, allows the author to specify the correct answer and synonyms for any of the words or phrases in the answer. Where markup of specific spelling is desired it is only necessary to include the command &M in the BASIC program.

These commands make it possible to rapidly develop CAI programs, in BASIC on an Apple II personal computer, using the most advanced answer judging algorithms available on any computer system at this time. In addition to answer judging, this authoring system (ref. 12) provides easy ways for student users and authors to enter upper and lower case, superscripts and subscripts, accents, and to use alternate fonts.

Where large numbers of students are using a CAI system it is sometimes desirable to avoid handling of individual diskettes by having each microcomputer connected to a single large disk system in a cluster system. One such system is the PLATO cluster system which is under development at the University of Illinois. In this system microprocessor based terminals (refs. 13, 14) are connected to a Motorola 68000 based computer with a 32 megabyte disk. The code necessary to support the lesson is automatically transferred, at 2 megabits per second, from the cluster computer to the terminal where it is executed locally. This allows the use of programs whose length is not limited by the size of the memory in the terminal. These microcomputer based clusters provide all of the communications and central data base features associated with current large computer systems.

The rapidly growing capabilities of low-cost microcomputers provide a means of delivering computer-assisted instruction at the personal level. Clusters of such system offer a way to provide individualized instruction in classrooms. To take full advantage of these hardware systems requires the further development of software tools to facilitate the writing of instructional programs.

REFERENCES

1 L. M. Jones, D. Kane, B. A. Sherwood and R. A. Avner, American Journal of Physics (IN PRESS) (1982).

2 C. Moore, S. Smith, and R.A. Avner, J. Chemical Education 57 (1980) 196

3 R. Chabay and S. Smith, J. Chemical Education 54 (1977) 745

4 P. Tenczar, Proceedings of the International Conference on Cybernetics and Society 443 (1981)

5 PLATO is a trademark of the Control Data Corporation. The PLATO system was developed at the University of Illinois and at the Control Data Corporation.

6 S. Smith and B. Sherwood, Science 192 (1976) 344

7 R. Chabay and S. Smith, General Chemistry Diskettes for the Apple II, COMPress, P.O. Box 102, Wentworth, N.H., 03282

8 Apple is a trademark of the Apple Computer Company.

9 S. Smith, Introduction to Organic Chemistry, Diskettes for the Apple II, COMPress, P.O. Box 102, Wentworth, N.H., 03282

10 S. Smith, J. Pochan, H. Gibson, Polymer Chemistry, Diskettes for the Apple II, American Chemical Society, 1155 Sixteenth Street, N.W. Washington, D.C. 20036

11 P. Tenczar and W. Golden, Spelling, Word and Concept Recognition, CERL report X-35, University of Illinois (1968).

12 P. Tenczar, S. Smith, and R. A. Avner, Diskettes for the Apple II, COMPress, P.O. Box 102, Wentworth, N.H. 03282

13 J. Stifle, S. Smith and D. Andersen, Proceedings Association for the Development of Computer-Based Instructional Systems 3 (1979) 1027

14 S. Smith, Microprocessor Delivery of PLATO Instructional Material, Personal Computers in Chemistry, Peter Lykos, Ed., John Wiley and Sons, New York (1981) 177.

S.R. Heller and R. Potenzone, Jr. (Editors), *Computer Applications in Chemistry*
© 1983 Elsevier Science Publishers B.V., Amsterdam — Printed in The Netherlands

LOCAL AREA NETWORKS, FOURTH GENERATION LANGUAGES AND MICROELECTRONICS IN THE RESEARCH LABORATORY

RAYMOND DESSY
and members of the Laboratory Automation and Instrument Design Group: MARK THOMPSON, IAN CHAPPLE, JAMES CURRIE, MARK WINGERD, JIM BERGQUIST, STEVE DUEBALL

Virginia Polytechnic Institute and State University, Chemistry Department, Blacksburg, Virginia 24061

INTRODUCTION

The advent of large scale microelectronics has led to a major change in the research laboratory. A decade ago it was possible to place four discrete circuit elements in one square centimeter of space. Today it is possible to place nearly half a million in that same space! By mid 1980 printed circuit board interconnections will routinely be at submicron level widths giving rise to fabrication elements smaller than a T4 bacteriophage. Multilayer ICs will no longer obey the rules of classical physics. Quantum mechanical descriptions will be needed due to interaction of small elements in close proximity. The world of chemical research will be irrevocably altered.

One such effect will be a continuation of the already prevalent incorporation of very powerful central processors into our analytical equipment for instrument self-calibration, data collection, data manipulation and report generation. There will be a continuation of the flight from large main-frame interaction toward the personal computer.

This flight is understandable. It is part of a technological revolution. The technology part of this revolution results from the rapidly increasing VLSI capabilities. Cost of fabrication, the urge for profit, and spiraling competition encourage manufacturers to produce small, extremely capable computer systems. The revolution itself is just that - a revolt of the masses against the tyranny imposed by large systems, their arcane language and tongue, formidable power structure and user unfriendliness.

LOCAL AREA NETWORKS

In industrial and governmental laboratories throughout the country one sees a common scenario - the power struggle and class lines drawn in the contention over who is to automate what in the laboratory and how it will be accomplished. One is reminded of the history of the church or perhaps the question of States rights vs. Federal control.

Perhaps an allegory less fraught with emotion represents the central corporate or laboratory computing center as a "made horse", well trained and capable of responding to the commands and aids of an experienced rider or trainer. Its repertoire is not easily changed nor can it tolerate the casual rider. A horse bred and trained for the track will not perform well in the three day event. Yet many propose the proper solution to laboratory automation is a single large computer facility.

In the face of such attitudes the laboratory worker often goes to the foal yard and purchases a young animal. It is manageable or at least small enough to give the sense of mastery to its new owner. But it does need training and during that period it will consume a large amount of feed; just as microcomputers and personal computers, although touchable and apparently controllable, consume software voraciously. Both foal and the microcomputer may eventually prove unsuited to the task originally envisaged or the task as it eventually evolves. The owner may find training and software writing involve skills that he does not really possess. A good chemist may be turned into a second class systems programmer.

This is not to decry the use of small personal or dedicated computers. It merely urges that the capabilities of the computer selected be suitable for the proposed task; a computer which can be integrated into a synergistic network with communication between its members. This will allow software development to occur at multiple levels within the interacting network. Where proper hardware, software tools and personnel are available the cost of development is reduced and the quality is increased. This is the first element in the title - Local Area Networks.

There is obvious evidence in those labs where a true cost

accounting is made that hardware is an insignificant factor in laboratory automation. In the laboratories of the Laboratory Automation and Instrument Design group at Virginia Tech, where 16 DEC PDP-11 family computers are interconnected in a net, the cost of the CPUs represents about 4% of the investment. Terminals represent 8% of the total cost and this is rising; peripherals comprise about 20%. All the rest is software.

FOURTH GENERATION LANGUAGES

Extensions have been added to the second generation languages of Fortran and BASIC like additions to houses as the family grows, and with the same consequences. We seem not to have recognized that there is a difference between languages intended for APPLICATIONS, and languages intended for what traditionally has been called PROGRAMMING. Many fourth generation languages are being developed which are ideal for application orientation at the laboratory level. These languages have the attribute of being extensible. They allow the definition of new words intended to implement specific functions, words that are human oriented. They are the same words used to communicate with fellow scientists and technicians. These words may be strung together just as we use words to create sentences and with the same effect and simplicity.

Such languages are the pathway to new instrument design in which the classic keyboard approach is replaced by a push-button approach with keys that are pressed in any desired sequence to create an applications command sequence. The Hewlett-Packard optical multichannel analyzer and the Perkin-Elmer CRT labeling of special function keys are examples that instrument manufacturers realize the tragedy conveyed when one sees a hand held calculator next to a computer terminal. It should not be harder to add 4 and 3 on a computer terminal than it is to perform the same function on a calculator. Experiments can and should be just as easy to control. Thus the second element in the title – Fourth Generation Languages.

MICROELECTRONICS IN THE RESEARCH LABORATORY

Finally, one of the most important offshoots of the microcomputer revolution for the chemist will be the use of microelectronic devices as transducers and the computer front-end processors. Thus the third element in the title – Microelectronics in the Laboratory.

A NETWORK EXAMPLE

The network in our laboratory is shown in Figure 1. Although specific to our needs it allows a philosophical discussion on general applicability.

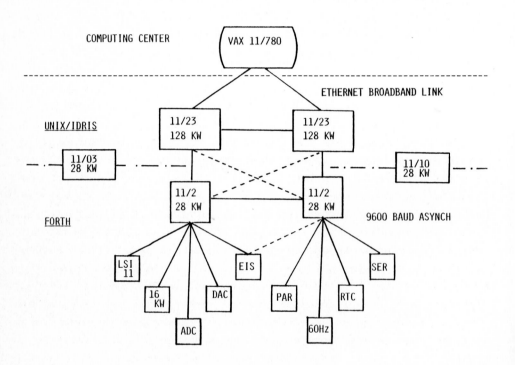

Figure 1.

The numerous small microcomputers are based on DEC LSI-11 CPUs with 16 KW of memory, equipped with serial, parallel, ADC and DAC interfaces. All real-time programs are executed at the LSI-11. Each research station is equipped with enough peripheral support (disk, plotter, graphics, etc.) to serve the immediate needs of the experimenter. The satellites are attached to two host computers where other non-immediate resources are available. Program preparation is done at these identical hosts. They are DEC LSI-11/2 CPUs with 28 KW of memory, supporting a FORTH, Inc. multiuser, multitasking version of FORTH. Adequate disk, graphics and printer/plotter equipment assures that program preparation is well supported. The user's programs and a FORTH nucleus are down-line loaded electrically from the host to each satellite when it is brought-up for experiment control. Adequate buffering at the satellite allows them to acquire data in real-time and survive minor down-times on the host to which it is attached. All data to be archived is returned to the host where it is stored on either real or virtual disk volumes on individual user surfaces.

The network communication at this level involves 9600 baud, RS-232 transmission with ASCII and transparent binary capabilities. A positive acknowledgment, check-sum and parity protocol leads to requests for re-transmission at either end on receipt of corrupted information. The net at this level is bilaterally symmetrical with two identical hosts. This allows half the net to survive during a repair session and also allows one to experiment with the operating system without completely disrupting the laboratory's operation.

Above this level, again in a bilaterally symmetrical arrangement, are two DEC PDP-11/23 CPUs with 128 KW of memory and large rigid disks. With microcoded implementation of 64-bit float arithmetic these processors are capable of data correlation and statistical manipulation. These tasks would be prohibited on the slower host processors beneath them by real-time constraints imposed by the human users and hardware peripherals. At this level a UNIX, or UNIX-like, operating system is used. The C language, commonly supported at this level, is pointer oriented with pipe-lining capabilities. It complements FORTH at the lower level. Although not really oriented toward real-time it does have powers and extensions FORTH does not have, yet is very similar in philosophy and implementation.

Each of the four hosts (two levels of two) can intercommunicate with each other. Full transparent terminal capabilities are being built into the system and currently exist two levels deep.

Above the dual PDP-11/23s an interconnect to the campus facilities exists over a broadband coaxial link. The Sytek 20/100

Packet Communication Unit at our remote site uses mid-split CATV technology to communicate with the head-end, which links with a data switch to allow interconnection with one of several VAX facilities on campus or with the IBM central computing facility. The ultimate goal is to have a campus wide net supported by a CSMA/CA protocol.

Increasingly, one sees chemical laboratories installing some type of baseband or broadband CMSA/CA local area network. When the number of units involved is small, point-to-point networks typified by the low end of our complete network are traditional, easy and inexpensive to install. As the number of installed stations rises cost and inconvenience also rise. A multiple-access, single coaxial wire transmission medium is preferred. Either time-division multiplexing (Ethernet, baseband) or frequency division multiplexing (broadband) may be used to increase the carrying capacity of that medium. Baseband is less costly to install, but can carry a limited number of users. Broadband can handle 10's of thousands of low speed device communication packets concurrently, but the technology is more expensive. In both cases IEEE Standards have yet to be determined. Despite that, many organizations are already installing the required communication channel. They feel that no alternative exists. The cost of point-to-point interconnect is prohibitive both in cost of cable and the space in existing steam tunnels and conduits for the cable to be installed.

The languages FORTH and C lend themselves naturally to the environment described. Network protocols have been either easy to write or easy to acquire. This is important even in friendly environments. Our broadband link is a collaborative effort between the Computing Center and our laboratory. The dashed line in Figure 1 delineates the ownership and responsibility boundary that should exist somewhere in any network. Below the line our laboratory decides what equipment is needed, how it is to be deployed, and how experiments are to be conducted. Above the line the domain is owned, administered and paid for through the Computing Center. The laboratory researcher has no need or justification to learn an excessive amount about how the Computing Center operates. It is a service. That service should not dictate how the laboratory is run. Thus the need for the dashed line. However, human and machine communication across that line is vital.

To our laboratory the network provides a tool just like spectrometers and reactors. At the lowest end are the real-time acquisition devices programmed by their respective users in a user-cordial application's oriented language. Storage of data and program requires a slightly larger computer which primarily serves as a disk controller and programming station. Above this level are processors with good math and storage facilities which can perform the statistical and arithmetic manipulations needed as well as archive data. Finally, for very large mathematical requirements and larger database management functions direct access to the campus facilities exist.

To think of these facilities being fused into a single system is to deny that form follows function or vice versa. Writing real-time code to handle all contention problems on a large multiuser, multitasking system is not trivial. Debugging it either requires a cumbersome operating system to protect the other users or risks the credibility of the system as it crashes periodically during the early iterative attempts to locate errors. Likewise, it is not the domain of small computers with limited resources to run math and data management programs. Yet a system in which all transfer between small machine and large machine is manual is equally unthinkable. The network is the only rational way to achieve true laboratory automation.

MICROELECTRONIC DEVELOPMENTS

The network is used extensively to aid in the development of microelectronic transducers and front-end processing devices. A few typical examples will be described.

Linear Array Detectors. Linear array silicon photodiode detectors based on charge-transfer devices are now commonly used as the basis for dispersive spectrometers in which no moving parts are required. The spectrum is dispersed by holographically engraved concave gratings onto the flat field of the detector. This may contain from 512 to 1024 detectors, all encased in a small one inch long IC. A dedicated computer is required for control and data acquisition. However, communication to a larger computer is necessary for data manipulation such as spectral subtraction, three dimensional hidden line plots, etc.

Spectrometers also based on silicon photodiodes may be made with discrete photodiodes covered by small 10 nm interference bandpass filters. Eleven of these can be placed on one IC carrier.Two of these can be combined to form a spectrophotofluorimeter. The programming of this device to handle excitation/emission matrix plots, or to manipulate the rheological equations to convert scatter at 5 wavelengths to particle size cannot be performed on a small machine. Yet that is exactly the type of CPU required by a portable field instrument. The network allows program preparation on one machine, followed by down-line loading/PROM burning for a simpler target processor.

Linear Array photodiodes can be used to receive the dispersed output of a Michelson interferometer. The dispersed interferogram allows each picture element (pixel) of the detector to represent information in a certain bandpass of the original spectrum. This considerably reduces the amount of data that need to be processed by a Fourier Transform algorithm to convert to the normal frequency domain. Solid water devices are currently being manufactured that can perform limited Fourier Transforms by a single high speed processor on a chip. Intended for telecommunication purposes, these front-end processing devices will certainly change the nature of FFT spectroscopy. The time frames involved are in the millisecond regime. The cost is insignificant.

Surface Acoustic Wave Detectors. Surface acoustic wave devices may be used to determine transitions occurring in polymer samples as a function of temperature. A small computer can handle the control of temperature from liquid nitrogen through 300 (degree) C along with data acquisition. It is not suitable for the archiving and correlation required by a record management system which will allow the data from hundreds of runs to be examined quickly and correlations plotted.

Nondispersive infrared equipment has traditionally handled the decomposition of data taken from a mixture into its component parts by a linear parameter estimation process involving matrix manipulations. Rewriting these matrices and handling them in a slightly different way indicates that correlation of information could yield the same concentration data. Devices built around charge transfer or SAW technology can do correlation in the analog domain quickly and inexpensively. Used in front of small microprocessors this equipment can not only perform the required math operations more quickly and at less cost, they can also reduce the size and sophistication of the microprocessor required. Such front-end processing also bypasses the programming problem created by digital manipulation of the data.

Microelectronics can do more than just implement computer technology. It can be the source of new detectors and math processing modules.

In the next decade solid state lasers, acoustic-optic and electro-optic dispersing devices and solid state electromagnetic detectors will change the world of spectroscopy. Solid state electrodes and sensing devices are changing the way electrochemistry can be performed. Coupled with the robotics and continuous flow analysis these new tools will give us the solution to the increasing problems being thrust upon the analytical chemist.

The computer is expanding both our mind's capabilities and our senses.

S.R. Heller and R. Potenzone, Jr. (Editors), *Computer Applications in Chemistry*
© 1983 Elsevier Science Publishers B.V., Amsterdam — Printed in The Netherlands

COMPUTER NETWORKS FOR SCIENTIFIC APPLICATIONS: A STATUS REPORT

C. WILLIAM KERN
National Science Foundation, Washington, D. C. 20550

BACKGROUND

The merger of computer and telecommunications technology is leading to major changes in the ways scientists interact, both among themselves and with remote resources (ref. 1). The drivers of these changes are inter-related and many in number but are rooted primarily in advances in microelectronics and digital processing. The former has brought smaller, cheaper, faster, and more reliable computers and instruments to virtually every segment of the research and educational enterprise. The importance of developments in the latter field, particularly packet switching, perhaps have a lower profile outside of engineering but they are also revolutionizing the way scientists process their data and exchange information. Although the ultimate impact remains to be demonstrated completely, we see already significant benefits to scientists who have had access to a digitized, computerized telecommunications world. The key example is the way computer scientists developed ARPANET into the backbone of a highly interactive, distributed research environment.

Prior to 1970, most interactive communication networks used circuit switching in which the transmission bandwidth is preallocated either permanently (e.g., radio) or for a fixed duration (e.g., telephone). The telegraph system, an exception, scheduled bandwidth dynamically one link at a time, thereby requiring manual sorting and routing decisions along the source to destination pathway. The introduction of computers into the communication field made it possible to divide the data being transmitted into small pieces, or packets, and to perform switching operations with incredible speed and reliability. This new technology, called packet switching (ref. 2), revolutionized the industry to the point that since 1978 virtually all new networks have been built around it. The economic lesson proved to be simple in substance (Fig.1): use packet switching when computer hardware costs fall below bandwidth costs.

Advances in related fields--fiber optics, lasers, and satellites--combined with packet switching and user demand appear to be leading us to a world-wide, point-to-point digital network with integrated multi-media services.

The purpose of this article is to survey briefly the current status of computer networks and to suggest factors that must be considered for application of the underlying technology to scientific disciplines such as chemical research and education. For greater detail see the references cited below.

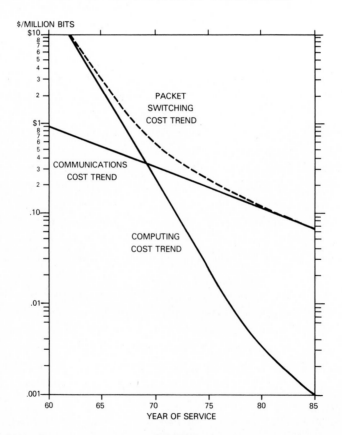

**INCREMENTAL COST OF MOVING 1M BITS (1 KILOPACKET)
ON A NATIONWIDE NETWORK**

Fig. 1. Cost Performance Trends for Packet Switching.

COMPUTER NETWORKS

By the term "computer network", we shall mean an interconnected set of autonomous computers. The computers can be the same or different; they can be located in the same building or on the same campus (local network); they can be distributed throughout the nation or around the world (long-haul network); they can be connected by cables, optical fibers, microwave circuits, lasers, or earth satellites. The purpose of a computer network is to enable information exchange among computers, whatever the interconnection distances and media.

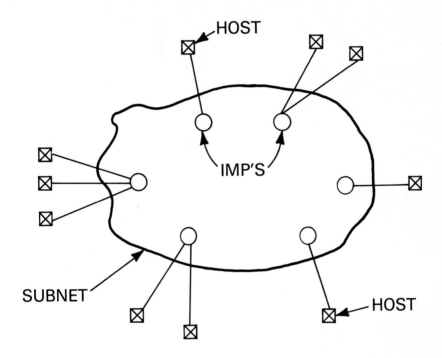

Fig. 2. Schematic Relation between Hosts (X), IMPS (0), and
the Communication Subnet.

Network Terminology (ref. 3)

Following ARPANET terminology, we shall refer to the computers connected to a net as "hosts". The hosts run user services or application programs. Each host or group of hosts is part of the network in the operational sense that information

can be transmitted from host to host via "interface message
processors (IMP's)". The IMP's are components of the
communication subnet (Fig. 2). Packets are sent from source to
destination IMP via intermediate IMP's which are instructed to
store and forward the packets as soon as a communication channel
(e.g., high speed telephone line) is available. The channels are
linked according to one of several topologies (Fig. 3).

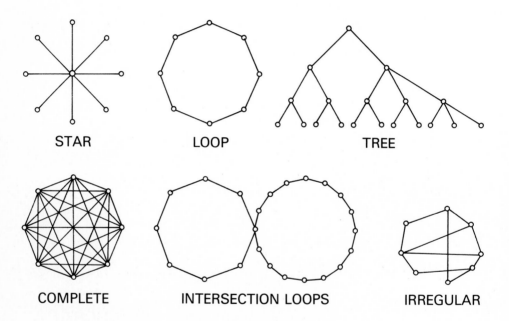

Fig. 3. Examples of Network Topologies.

So far, we have seen that a computer network consists of
service elements (i.e., hosts) and communication elements (i.e.,
IMP's and channels). A protocol is a format and a set of
conventions for the exchange of information. To reduce the
complexity of network design, a hierarchical, layered
structuring of protocols is used to facilitate the
implementation of services such as message services and file
transfer services. Although not universally applied, the
International Standards Organization (ISO) has developed a seven
layer model (Fig. 4) in which lower layers support operations

and services of higher ones. Links between hosts occur at most
layers which transmit information according to common protocols.
In this manner, the functions of each layer can be organized and
programmed independently in a variety of systems thereby
facilitating communication between diverse types of computers.
The term network architecture includes specification of the set
of layers and their concomitant protocols. Although each
protocol level requires software to be written and installed in
the hosts and packet switches, the scientific user is usually
less concerned with the underlying protocols and more with
available services.

LAYER NUMBER	ISO	ARPANET
7	APPLICATION (USER PROCESSES)	APPLICATION
6	PRESENTATION (EDITING)	TELNET, FTP, SMTP
5	SESSION (SESSIONS)	(NONE)
4	TRANSPORT (MESSAGES)	TRANSPORT (TCP)
3	NETWORK (PACKETS)	INTERNET (IP) NETWORK (HOST/IMP)
2	DATA LINK (FRAMES)	LINK (HDH)
1	PHYSICAL (BITS)	PHYSICAL

Fig. 4. Approximate Correspondence between the ISO Seven Layer
        and ARPANET Protocol Models. Telnet = Virtual Terminal
        Protocol; FTP = File Transfer Protocol; SMTP = Simple
        Mail Transport Protocol; TCP = Transmission Control
        Protocol; IP = Internet Protocol; Host/IMP = BBN 1822;
        HDH = HDLC/Distant Host Protocol.

Network Services

For scientific users, computer network services (ref. 4) normally fall into three basic categories:

Messaging. (ref. 5) The most elementary application of a network, and yet the most valuable judging by the volume of traffic, involves the transmission of messages between users with accounts on common or different hosts. The "messages" may be short textual material containing, let us say, up to 2,000 characters or they may be much longer files containing documents, data, instrument or design specifications. Messaging software systems contain a multiplicity of features beyond simply sending and receiving a single message. They allow distribution simultaneously to a list of recipients or to a bulletin board, which is effectively an electronic mailbox that can be read by all the users belonging to the network community; they support direct communication, as in a face-to-face meeting, by displaying simultaneously the latest messages from all participants engaged in the conference, debate, manuscript preparation, or whatever other form the communication may take; some systems also provide indexing, editing, and archiving features, as well as simplified commands for answering or forwarding messages. In short, "messaging" presents the user with powerful tools and constitutes the majority of net use: large file transfers are responsible for the most of the bits, and interactive use for most of the packets, flowing over the ARPANET, for example.

Remote-System Accessing. Perhaps the simplest demonstration of a computer communication network is to allow two hosts to act as terminals for one another. This application is similar to using a dial-up telephone port to access a computer remotely but with the difference that information flow is much faster (9600 vs 1200 bits per second). A more sophisticated type of application, and indeed one still being developed at the forefront of computer science research, involves a subarea of distributed computing called interprocess communication. For example, the network may allow different parts of an applications program to be performed on different computers at different physical locations. The network system handles all administrative and data management tasks, from instructing other nodes to execute a given process, returning the answers error-free, and combining them in the proper sequence to produce the final results. In such applications, where all resources in the network are available to each host as if a local homogeneous multiprocessor system were being utilized, the distributed nature of the actual execution may be transparent to the user, except when it comes time to determine the overhead cost of using the network!

Non-Textual Embedding. The services outlined above can be extended by recognizing that the IMP and host operations occur as digital processes. Thus, by converting voice signals and facsimile images to digital form, transmitting those data over the network, and then reconstructing them at the receiving host, it is possible to perform multi-media messaging. Because these services are data intensive (1 sec of speech = 100,000 bits, 1 color image = 100,000,000 bits), high band widths and compression techniques are the keys to practical implementation over long-haul networks. So far, research has led to speech compression in the 1200 to 2400 and color compression in the 128,000 to 256,000 bits per second ranges, respectively.

Examples of Long-Haul Networks

The first comprehensive documentation of packet-switched networks was prepared by Baran in 1964 who at that time was working for the Rand Corporation under contract to the U.S. Air Force. Baran proposed a highly distributed ("survivable") computer network to provide for all military messaging including encrypted data and voice communication. In 1969 the (Defense) Advanced Research Projects Agency (DARPA) initiated a program plan for what was to become ARPANET (Fig. 5). This included the acquisition of network circuits and selection of Bolt Beranek and Newman, Inc. (BBN) as the systems contractor for the design of the subnetwork and IMP's. Further details about the history of ARPANET can be found in reference 6.

In about the same year (1969), Tymshare Corporation began installing a store-and-forward network based on minicomputers to connect user terminals to its central mainframes. This led with FCC approval in 1977 to TYMNET which is now the largest public packet switched network in the United States. TELENET offers similar services and was the first U.S. public net (1975) to adopt international connection standards (x.25) and to connect

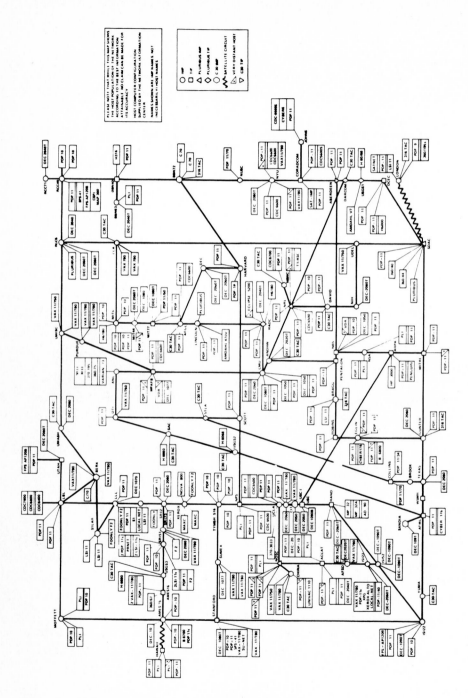

Fig. 5. ARPANET Logical Map, March 1982.

with commercial networks abroad (Fig. 6). Links between packet
switches   use   high   bandwidth   terrestrial   or   satellite
communication channels, with bit rates commencing at 4,800 bits
per second (bps). User terminal-to-host links have transmission
speeds between 300 and 9600 bps. The salient features of
ARPANET, TYMNET, and TELENET are given in Table 1.

INTERNATIONAL ACCESS TO TELENET

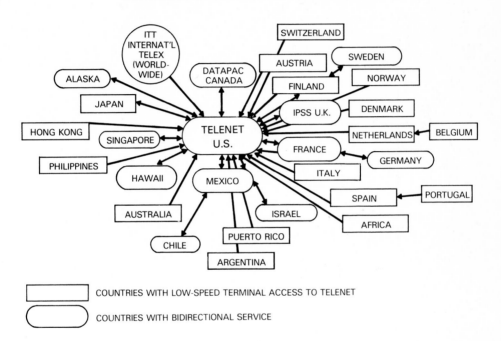

Fig. 6. Telenet Geographical Map, February 1982.

    Most large corporations have their own private data networks
and even sell them to customers (IBM/SNA, DEC/DECNET, ATT/ACS);
computer communication networks are also developed with public
funds.
    The U.S. Department of Defense   (DOD) began devoting
substantial   resources   in   1977   to   the   development   of
high-performance automatic digital network called AUTODIN II.
Following several major delays in the project, which was
designed to allow military users to send classified data
worldwide without special coding procedures, the Deputy

TABLE 1. INTERFACE, OPERATION, AND SERVICE FEATURES OF ARPANET, TYMNET, AND TELENET

INTERFACES	OPERATIONS	SERVICES
ARPANET		
DATAGRAM (UP TO 8063 BITS & HEADER) DESTINATION ADDRESS ON ENTRY SOURCE ADDRESS ON EXIT	PACKET SWITCHED, STORE-AND-FORWARD (LAND LINES POINT-TO-POINT SATELLITE CIRCUITS) UP TO 1008 BIT PACKET & HEADER ADAPTIVE ROUTING FLOW/CONGESTION CONTROL 95 PACKET SWITCHING NODES MODERATE BANDWIDTH BACKBONE LINKS (50 KB/S) (SOME 96. KB/S CIRCUITS, SOME 230.4 KB/S) HIGHSPEED LOCAL HOST/PACKET SWITCH CONNECTION (100–250 KB/S) LOWER SPEED REMOTE HOST/PACKET SWITCH CONNECTION (9.6–50 KB/S)	CHOICE OF SEQUENCING/NON-SEQUENCING DELIVERY CONFIRMATION < 500 MS. DELAY POINT-TO-POINT ADDRESSING END/END SECURITY (ADD-ON, NOT INTERNAL SERVICE) LOW UNDETECTED BIT ERROR RATE ($<10^{-12}$) CHARACTER MODE TERMINAL ACCESS ("NETWORK VIRTUAL TERMINAL")
TYMNET		
VIRTUAL CIRCUIT, CHARACTER MODE EXPLICIT SOURCE, IMPLICIT DESTINATION ADDRESS	STORE AND FORWARD "FRAME" SWITCHING OVER POINT-TO-POINT CIRCUITS VIRTUAL CIRCUIT ROUTING ESTABLISHED AT INITIATION LINK BY LINK FLOW CONTROL LINK BANDWIDTHS (2.4–9.6 KBLS)	TERMINAL TO SERVER HOST CHARACTER MODE VIRTUAL CIRCUITS POINT-TO-POINT SERVICE ONLY LOW UNDETECTED BIT ERROR RATE 10^{-8} EXPLICIT HOST/HOST SERVICE
TELENET		
VIRTUAL CIRCUIT, UP TO 8192 BITS & HEADER SOURCE AND DESTINATION ADDRESS VARIABLE LENGTH/VIRTUAL CIRCUIT ID	STORE AND FORWARD PACKET SWITCHING (OVER POINT-TO-POINT CIRCUITS) UP TO 1024 BIT PACKET & HEADER ADAPTIVE ROUTING FLOW/CONGESTION CONTROL PER SUBSCRIBER VIRTUAL CIRCUIT > 80 PACKET SWITCHING NODES BACKBONE LINK BANDWIDTHS (MOSTLY 9.6 KB/S) HOST INTERCONNECTION LINK BANDWIDTHS (300–50K BPS)	VIRTUAL CIRCUIT ONLY — SEQUENCING, DELIVERY/NON-DELIVERY CONFIRMATION 500 MS DELAY POINT-TO-POINT ADDRESSING LOW UNDETECTED BIT ERROR RATE ($<10^{-12}$) CHARACTER MODE TERMINAL ACCESS (VIA "PAD")

Secretary of Defense directed in April 1982 that AUTODIN II be
terminated after extensive reviews of technical, managerial, and
cost issues. Instead, it was determined that the ARPANET
approach could and should be expanded and combined by 1986 with
the complementary military network development efforts indicated
in Fig. 7. Note that the present ARPANET is to be split into an
R&D Network and a Defense Data Network (DDN).

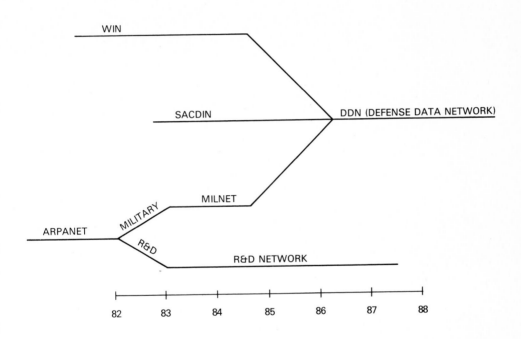

Fig. 7. Projection of DOD Network Development.

    The DOD has also provided important leadership in developing
ways to handle traffic between hosts on **different** computer
networks. This type of internetwork connection problem poses a
variety of new technical and jurisdictional issues; these
include functionality and operational issues of the internet
"gateways", addressing users on the different networks, routing
decisions, and allocating user charges. The situation becomes
particularly interesting when the component nets have different
protocol hierarchies. For example, the International Standards

Organization (ISO) and ARPANET protocols present a mismatch
(Fig. 4) in the sense that the former model explicitly lacks so
far an "internet" protocol layer.

Fig. 8. CSNET Geographic Map, July 1982.

A facility called CSNET (Computer Science Research Network)
(ref. 7), being developed with the support of the National
Science Foundation, is addressing this issue by linking a
variety of networks, including ARPANET, TELENET, and PHONENET, a
telephone-based relay network. Figure 8 shows that CSNET hosts
may be categorized as ARPANET hosts, TELENET hosts, or PHONENET
hosts, depending on the network to which they are primarily
connected. There is also a Public Host, which is a CSNET VAX 750
available to users who do not have access to any other CSNET
host. A CSNET Service Host is designed to provide network
services including nameserver directory facilities, accounting
of network traffic, and network addressing and routing tables.
The PHONENET relays are VAX 750's that relay files to and from
PHONENET hosts which are required to implement a mail transport

system called a Multichannel Memorandum Distribution Facility
(MMDF) (ref. 8). Users of CSNET hosts can communicate with one
another regardless of the component network to which their hosts
are connected. Further details about CSNET can be found in
(refs. 7 and 8).

The CSNET project is part of a cooperative effort to
establish a network facility among computer science research
groups in universities, industry, and government. Basic
messaging and accessing to remote systems will be provided
initially as communication services. After the five-year, $5M
development period is completed in 1986, costs will depend on
the services utilized in such a manner as to make CSNET
self-sustaining. The CSNET facility will be available to all
U.S. groups engaged in a computer science research: this is the
primary membership criterion.

Another significant example of a computer network system,
serving a community of scientists with common interests is
PROPHET, developed at BBN over a period of more than a dozen
years under sponsorship from the National Institutes of Health.
PROPHET provides a system of maintaining files of standard
molecular structures, experimental data, etc., to research
biochemists, pharmacologists, and crystallographers. PROPHET
also enables these scientists, now numbering over 700 users at
40 sites throughout the United States, to perform certain types
of computations, including regression analyses of data,
reformatting data files, and preparing graphical output. Since
users may share data including textual information, the network
environment fosters new collaboration among users.

The National Institutes of Health have also supported since
1974 a network called SUMEX-AIM (Stanford University Medical
Experimental Computer--Artificial Intelligence in Medicine)
which is concerned with providing an advanced computational
environment for medical scientists distributed throughout the
United States. Spin-offs to modern genetics (GENET) and
cognitive scientists (COGNET) are under various stages of
discussion.

PROPHET and SUMEX-AIM are examples of star networks (Fig. 1)
in which a single host computer is accessed by geographically
dispersed terminals. Terminal access to the host is by one or
more communication subnetworks, for example TELENET.

As an example of a long-haul network in a different field, we cite the Magnetic Fusion Energy Network (MFENET) which has been supported by the U.S. Department of Energy. MFENET has over 40 nodes including three mainframe computers which consist of two Cray machines and one CDC-7600. There is a message system that sends and receives messages over four satellite links and terrestrial lines. In addition to messages, MFENET is designed to transmit large files of data over the network communication system.

Other groups of physical scientists, for example those at the National Center for Atmospheric Research, are also considering the construction of a national packet-switched network system which would interface with their computational facilities in Boulder, Colorado.

COMPUTER NETWORK DESIGN FACTORS

From the examples in Section 2, it can be demonstrated that scientists can gain unique benefits from long-haul computer-based networks if a number of conditions are met. First a group of users must be identified who want to pursue common professional interests through the network. Second, these users must have financial resources to construct the type of system that meets their requirements. Third the design objectives of the network must be determined clearly at the outset. Finally, steps must be taken to ensure that the network, once in place, is "friendly", that is customers are equipped to gain advantages by using the net and are satisfied with its performance.

Although construction of the network is essentially a technical task, it is the users who must make the fundamental design choices some of which are described below.

Design Objectives (ref. 9)

Perhaps performance levels are the single most critical element in the design of computer communication systems. These can be subdivided into speed and service levels, as indicated in Table 2. Speed can be subdivided further into delay and throughput. Delay is measured by response time, meaning the time between transmission and delivery of the first bit of information. This is determined by the length and bandwidth of the circuit as well as by any processing delays at packet switches or queueing delays at the destination host. Throughput, the effective bit rate of information per second flowing through the system, is determined by the slowest component in the total system, whether it be the transmission medium or the nodal processing equipment. Low delay is needed for most interactive applications of the network and high throughput for large volume messaging.

TABLE 2. Performance Factors in Computer Network Design.

FACTOR	PARAMETER	MEASURE
SPEED	DELAY	AVERAGE RESPONSE TIME
	THROUGHPUT	PEAK TRAFFIC LEVEL
SERVICE	AVAILABILITY	PERCENT DOWN TIME
	DATA INTEGRITY	BIT ERROR RATE
	MESSAGE INTEGRITY	LOSS RATE
	SECURITY	MISDELIVERY RATE

Another set of design factors implied by network objectives concerns data representations, data transfer, and grades of service. For example, will the communication system support only one datum type (ASCII) or will it support unrestricted binary data? Will the data format fields be fixed or free? Should messages be delivered in a completely transparent manner or should the system provide certain amount of data with code conversion? And what about the level of error control along the communications path? Should the grade of service be first-come, first-served or should the system recognize a priority hierarchy, based for example on ability to pay for services. These and other questions regarding the functionality of the network depend on its basic purposes. Having decided on the performance and functionality levels of the communications system, one must then decide which applications will be supported, both at the outset and in future years. This, in turn, will help to identify specific hardware, software, protocols, and interfaces.

A variety of cost factors must also be examined. Will the network cost less than the system it replaces? Will the system

be self-supporting in terms of subscriber costs and/or
membership fees? Only when all of the absolute, relative, and
incremental costs are identified can the protocols optimize
cost-effectiveness.

Another major area of choice involves transmission
facilities. First, the transmission equipment must be
appropriate in terms of frequency range; the higher the
frequency, the more bandwidth is available for communication.
Once the bandwidth is determined, one must decide next on the
transmission medium or media. A range of alternatives is given
in Table 3. Finally, the general type of transmission facility
must be addressed. For example, should the circuits be digital
or analog, dedicated or shared, point-to-point or multipoint?
What type of supplier should be approached, public or private?
The attributes of the various choices are summarized in Table 4.

TABLE 3. TRANSMISSION MEDIA CHARACTERISTICS

MEDIUM	CHARACTERISTICS
VALUE ADDED NETWORK	BROAD AVAILABILITY FLEXIBLE LEASED SERVICE USAGE-BASED CHARGES DISTANCE-INDEPENDENT CHARGES MAINTENANCE AND UPGRADING BY CARRIER
COMMON CARRIER CIRCUIT	READILY AVAILABLE LEASED SERVICE MAINTENANCE BY CARRIER EASY RECONFIGURATION PRACTICAL OVER MEDIUM TO LARGE DISTANCE VARIETY OF SPEEDS
OPTICAL FIBER	HIGH BANDWIDTH SMALL DIAMETER WEIGHT NO CROSSTALK NO RFI POTENTIAL FOR LOW COST VERY LOW ERROR RATES
SATELLITE	HIGH BANDWIDTH AT LOW COST LARGE DELAY DISTANCE INDEPENDENCE BROADCAST MEDIUM LOW INCREMENTAL COST

TABLE 4. TRANSMISSION FACILITY TRADEOFFS

PUBLIC VS. PRIVATE FACILITIES

PUBLIC
- LONG DISTANCE LINKS
- MINIMAL INSTALLATION COST
- LEASED SERVICE
- MAINTENANCE BY CARRIER
- LIMITED OFFERINGS
- EASY RECONFIGURATION

PRIVATE
- SHORT DISTANCE LINKS
- POSSIBLE USE OF INEXPENSIVE MEDIA
- USER INSTALLATION
- LEASE/PURCHASE EQUIPMENT
- USER MAINTENANCE
- WIDE VARIETY OF OPTIONS
- MORE COMMITMENT TO EQUIPMENT

ANALOG VS. DIGITAL FACILITIES

ANALOG
- REQUIRES MODEMS
- MOST WIDELY AVAILABLE OFFERINGS
- VARIETY OF CIRCUIT SPEEDS
- ACCEPTABLE ERROR CHARACTERISTICS
- TELEPHONE SERVICE

DIGITAL
- NO MODEMS REQUIRED
- MORE COST EFFECTIVE
- LIMITED BUT USEFUL SET OF SPEEDS
- BETTER ERROR CHARACTERISTICS
- DIRECTION OF FUTURE OFFERINGS

DEDICATED VS. SHARED FACILITIES

DEDICATED
- HIGH COST PER USER
- NO CONTENTION AMONG USERS
- LOW DELAY
- POOR UTILIZATION FOR BURSTY TRAFFIC
- SIMPLE

SHARED
- REDUCED COST PER USER
- USER BANDWIDTH DEPENDENT ON LOADING
- CONTENTION CAUSES QUEUEING DELAY
- GOOD UTILIZATION FOR BURSTY TRAFFIC
- ADDITIONAL HARDWARE FOR SHARING

POINT-TO-POINT VS. MULTIPOINT FACILITIES

POINT-TO-POINT
- POTENTIALLY EXPENSIVE FOR MANY USERS
- SIMPLE CONTROL
- NO WAITING FOR OTHER SUBSCRIBERS

MULTIPOINT
- INEXPENSIVE COMMUNICATION BETWEEN SUBSCRIBERS
- MORE COMPLEX CONTROL
- QUEUEING DUE TO OTHER SUBSCRIBERS

SINGLE MEDIUM VS. MIXED MEDIA FACILITIES

SINGLE MEDIUM
- SIMPLE
- MAY BE MISMATCHED TO SOME TRAFFIC TYPES
- GOOD FOR UNIFORM TRAFFIC
- VULNERABLE
- MAY BE INFLEXIBLE FOR GROWTH

MIXED MEDIA
- MORE COMPLEX
- POTENTIALLY SUITABLE FOR ALL TRAFFIC TYPES
- GOOD FOR HETEROGENEOUS TRAFFIC
- LESS VULNERABLE
- MORE POSSIBILITIES FOR FLEXIBLE GROWTH

The choice of switching method is another major decision area in the design of a computer communication system. These are three possibilities: circuit, message, and packet switching. The basic question is: how can the users best share the transmission media? A circuit switched network provides service by setting up a dedicated physical path between two hosts. Although circuit switching is appropriate if the hosts need to communicate at a constant rate for a long time, it is inefficient for bursty traffic with high peak-to-average ratios. In message switching, the message is routed dynamically intact through the communication subnet. This requires provision for storage at intermediate packet switches and can lead to substantial delays. In packet switching, messages are sliced into packets which are routed independently and reassembled at the destination host. This method offers the most dynamic switching method since it makes effective use of circuit bandwidth and rapid computer switches. In Table 5, we compare these three switching methods.

TABLE 5. COMPARISON OF SWITCHING ATTRIBUTES

ATTRIBUTE	CIRCUIT SWITCHING	MESSAGE SWITCHING	PACKET SWITCHING
PHYSICAL CONNECTION	YES	NO	NO
REAL TIME	YES	NO	YES
DATA STORAGE	NO	YES	TEMPORARY
BLOCKING WITH OVERLOAD	YES	NO	YES
DELAYS WITH OVERLOAD	NO	YES	YES
ERROR CONTROL	NO	YES	PARTIAL
SPEED/CODE CONVERSION	NO	YES	YES
DELAYED DELIVERY	NO	YES	POSSIBLE
MULTIADDRESS	NO	YES	POSSIBLE

Another design factor concerns the topological layout of the system (Fig. 3). The basic problem is to find the minimum-cost network topology given various requirements, including user locations, traffic flow patterns, and performance characteristics (Table 2).

Selecting communication equipment for the network is also very important and sometimes difficult given today's marketplace. The basic classes of equipment required and their services are given in Table 6. One may also be concerned with satellite earth stations or internet gateway machines.

Another key issue is what protocols should be adopted inside the subnetwork? Recall that processes within hosts communicate with other hosts or terminals via computer protocols (Fig. 4). Thus, protocols are the core of a computer communication network. Protocols may also require that certain interface problems be addressed such as in the CSNET example.

TABLE 6. COMMUNICATIONS EQUIPMENT

HARDWARE	SERVICES
FRONT-END PROCESSORS	POLLING TERMINALS MESSAGE QUEUEING CODE CONVERSION
CONCENTRATORS	ADDRESSING BLOCKS OF DATA DATA COMPRESSION ERROR CONTROL/RETRANSMISSION
CIRCUIT SWITCHES	PATH ESTABLISHMENT CALL INITIATION/TERMINATION BILLING
MESSAGE SWITCHES	PRIORITY MULTIADDRESS MESSAGE RETRIEVAL
PACKET SWITCHES	INTERFACING ERROR & FLOW CONTROL SEQUENCING FAILSAFE OPERATIONS
NETWORK MONITOR CENTERS	LINE QUALITY MONITORING REMOTE TESTING & CONTROL OF SOFTWARE

CONCLUSIONS

For the most part, computer communication networks is less attributable to a single new technology but rather to the juxtaposition and hierarchical structuring of powerful, low-cost component systems. Moreover, internal design complexities are usually masked from the user who maintains control of the available services. For example, messages and data appear at the initiative of the scientist.

Most of the network components are likely to become available in an enhanced form during the remainder of this century. Costs, particularly for hardware, are likely to continue to decline, perhaps becoming almost "free" by today's performance standards. We should expect to see new public offerings of TELETEXT and VIDEOTEXT services (ref. 10) but these will probably be directed to the consumer (teleshopping, interactive games, how-to guides, home security), not to the scientific user requiring high-performance host-to-host services. "Scientific" networks will almost surely continue to be developed in a laissezfaire manner although some of them may be connected by internet gateways. Thus, it will be left to groups of individual scientists with common professional interests to take appropriate organizational initiatives if they wish to take full advantage of network technology. One might also expect, however, that the picture abroad will develop in a more organized, directed manner, where already one finds in place substantial national (e.g., DATAPAC/Canada) and international (e.g., EURONET) network structures and expertise.

ACKNOWLEDGMENTS

The author is grateful to his associates in the computer science research community who introduced him to this fascinating technology, particularly to those involved with the CSNET project. Professor L.H. Landweber and Dr. V.G. Cerf kindly read and made helpful comments on the manuscript. The material in the Design Objectives Section draws heavily on (ref.9) which, together with (ref. 3), are recommended highly as primers for the field of computer communication networks. Tables 1-6 have been largely reproduced, with permission, from the Introduction to Ref. 9, which was written by J. E. McQuillan and V.G. Cerf.

REFERENCES

1 For a recent survey, see *Science 215*, No. 4534, February 12,1982 issue, edited by P.H. Abelson and M. Dorfman.

2 L.G. Roberts, Proc. IEEE 66 (1978) 1307.

3 A. S. Tannebaum, *Computer Networks*, Prentice-Hall, Englewood Cliffs, N.J., 1981. Figures 1 - 2, p.8 and 1 - 3, p.9 are reprinted by permission of Prentice-Hall, Inc.".

4 An excellent discussion is given by A. Newell and R.F.Sproul, *Science* 215 (1982) 843.

5 R.E. Kahn, A. Vezza, A. D. Roth, eds., "Electronic Mail and Message Systems: Technical and Policy Perspectives", Proceedings of an AFIPS Workshop,(1980).

6 "A History of the ARPANET: The First Decade" Report No. 4799, (1981); prepared for DARPA by BBN.

7 L.H. Landweber and M. Solomon, "Use of Multiple Networks in CSNET", Proc. Compcon Conference, (1982).

8   D.   Crocker, E. Szurkowski, and D. Farber, "An Internetwork
Memo Distribution Capability - mmdf", Proceedings of the
Sixth Data Communications Symposium, (1979).

9   J.E. McQuillan and V. G. Cerf,   Editors. "A Practical View of
Computer Communication Protocols", IEEE Press, Report No. EHO
137-0 (1978).

10 J. Tydeman, et al.,   "Teletext and Videotext in the United
States", Institute for the Future, Menlo Park, CA, (1982).

S.R. Heller and R. Potenzone, Jr. (Editors), *Computer Applications in Chemistry*

COMPUTERIZED STANDARD REFERENCE DATA

BETTIJOYCE BREEN MOLINO
Office of Standard Reference Data, National Bureau of Standards
Washington, D.C. 20234

Much is being written in the current literature concerning the flow of information from the generator to the user. Traditionally, the channel for the flow of scientific and chemical information was based on primary publication in scientific journals, a transfer system which is "document oriented." Due to the volume and overlap of information in these journals, the need for abstract journals to provide reader guidance emerged about 75 years ago. However, the explosion of scientific and technical information over the past several decades has even taxed these abstract journals, and keeping abreast in his field has become nearly impossible for the scientist.

Thus, we face a need for better methods of transferring scientific and technical information from the originator to the user. Included in this general problem are such factors as shortening the time lag from generation of data to their incorporation into the working body of useful information for a specialist in a given field, better coordination and standardization of format and quality, and elimination of duplication. Facing the user of this information are such basic issues as where to find the data, are they complete, are they compatible with other sources, and how reliable are they.

The solution to this problem involves a change in the philosophy behind the flow of information. The document-oriented system is not sufficient; a single piece of scientific information should be the unit of flow into the distribution system. (ref. 1) This concept implies a condensing of the literature through organization, as well as a selection and evaluation process that assures reliability. By the term "evaluation of data," it is implied that published measurements of the quantity in question

have been carefully examined by experienced specialists in an
effort to select a recommended value and to give a statement
concerning the accuracy and/or reliability of this quantity. (ref.
2) The usefulness and the economic benefits of such evaluations
have already been demonstrated, as evidenced by the fact that many
compilations have already been produced throughout the world.

In order to integrate existing government-sponsored activities
and new programs into a coherent whole, the National Standard
Reference Data System (NSRDS) was established in 1963, with the
responsibility of administering that system being delegated to the
Office of Standard Reference Data (OSRD) at the National Bureau of
Standards (NBS). This system was strengthened in 1968 when PL
90-396, better known as the Standard Reference Data Act, was
signed into law, stating that "it is the policy of the Congress to
make critically evaluated data readily available to scientists,
engineers, and the general public . . . The Secretary (of
Commerce) is authorized and directed to provide or arrange for the
collection, compilation, critical evaluation, publication, and
dissemination of standard reference data." (ref. 3)

Today, the Office of Standard Reference Data administers one
of the largest data evaluation networks in the world. This
program, consisting of 22 data centers and numerous other data
evaluation projects, generates and disseminates critically
evaluated scientific and technical data, drawing upon scientific
expertise in Government, industry, and the academic community. The
extensive functions of the office include:

-allocating that part of the NBS budget which is spent on
  critical data evaluation, including that spent within the
  Bureau and that spent through contracts with outside groups
-monitoring all projects which the office supports
-managing the publication program of NSRDS
-providing information services
-maintaining close contact with other data compilation
  activities, both in the United States and abroad
-attempting to avoid duplication
-encouraging the coverage of all important technical areas
-engaging in research and development of techniques for the
  computerized handling of data and of applications for the
  automation of procedures concerning the activities of the data
  centers. (ref. 4)

Many of these functions, specifically those encompassing data evaluation processes and methods of dissemination, have been drastically altered in the last decade by advances in computing. (ref. 5) The 1970s saw a proliferation of online bibliographic files and, indeed Cuadra Associates has recently announced that there are now over 1,000 online databases, up 47 percent over one year ago. There is a continued growth in the diversity of subjects covered and, although to date emphasis has been placed on the bibliographic data, numeric data banks are beginning to follow suit. The Office of Standard Reference Data has cataloged numeric databases in a recent publication by Joseph Hilsenrath, NBS Technical Note 1122, <u>Summary of On-Line or Interactive Physico-Chemical Numerical Data Systems.</u> This publication gives a brief description of each of 53 interactive physicochemical numeric data systems, most of which are online on international computer networks. The systems are categorized as follows:

- those aiding in the identification of substances from spectroscopic data
- those providing thermodynamic and transport properties of pure components and mixtures
- those performing metallurgical calculations and drawing phase diagrams
- those producing complete tables of thermodynamic properties of individual substances
- those systems performing simulation, optimization, and design of chemical processes.

Information is given for each system, including the function, specific comments, the number of compounds in the database, the computer(s) and networks(s) on which the system is available, and the contact individual.

At the same time, OSRD has intensified its efforts to automate its data centers and to convert various "paper" databases to machine readable. In addition to the obvious day-to-day operational benefits to the data centers, such machine-readable files have the advantages of being able to be distributed in automated form (i.e., magnetic tape, cassette, floppy disk), via online systems, and/or in photocomposed form.

We will provide specific examples of computerized standard reference data and the methods of dissemination later in this report. First, it is important to recognize some of the reasons

that the development of online scientific numeric databases has
been so slow compared to the proliferation of bibliographic
databases. One fundamental difficulty lies in the representation
of chemical structure and nomenclature. In the type of scientific
data we are discussing, the numeric physical and chemical
properties are associated with specific substances, materials, and
mixtures. Any computerized system, therefore, must allow for all
the idiosyncrasies of nomenclature, including superscripts,
subscripts, upper- and lower-case Greek letters, common names, and
synonyms. The problems associated with nomenclature are compounded
when one considers the challenges involved with computerizing
chemical structures. To date, conventional database software does
not endorse such special needs. The software which has been
developed by OSRD and others is tailor-made and expensive.

A second problem with computerizing standard reference data
concerns the representation of the numeric values. The scientific
numbers, contrary to those used in business or economics, vary in
magnitude from very small to very large, requiring the use of
exponents. The number of significant figures is important and
needs to be conveyed with the number, along with the associated
error. A related issue to the representation of the numeric values
is the presentation of the data. Often scientific data are not
tables of numbers but, rather, models describing an infinite set
of conditions dependent on such parameters as temperature,
pressure, and concentration. The models provide predictive
capabilities as well as interpolation and extrapolation. Other
complications in presentation arise when the data require visual
display or graphic representation.

Finally, the computerization of scientific data requires the
objective judgment of the validity of the data by a scientific
expert. The National Standard Reference Data System has fulfilled
this need through its data center network, and the computerized
data banks it endorses are certainly evaluated data. The problems
involved encompass the level of effort and support needed to
capture the historical data while at the same time keeping current
with new data.

We have focused here on the problems involved with the data
themselves. In the computerization of scientific data, however,
there is a whole host of related problems concerned with the
computerization process - ensuring file maintenance and update,

using up-to-date and transportable computer technology, and providing user interface, training, and support.

Despite these areas of concern, the advantages of computerized scientific data are significant, and there is consequently a strong effort to develop in that direction. Computer files are easier to edit, update, and reformat. They can be more easily manipulated for scientific calculation, and more sophisticated searching is possible for data retrieval. Subsets in varied formats can be sent to other programs or even forwarded for photocomposition. Even the mere physical storage, back-up, and management of the scientific data are facilitated through computerization. Computerized data banks could never totally replace the published literature, but they can certainly be used to augment and enhance the end user's retrieval of his required information.

We have already discussed the important part the Office of Standard Reference Data plays in the critical step of data evaluation, in addition to the coordination and stimulation of data activities in the scientific disciplines and technical areas. Let us now look at some of the specific data files, which the Office of Standard Reference Data presently maintains in computerized versions, and examine the various means of disseminating them.

One of the earliest computerized data banks which the Office of Standard Reference Data made publicly available was the Mass Spectral Database, containing the electron ionization mass spectra for over 30,000 unique compounds. The data were collected from a variety of sources as part of a joint program of the National Institutes of Health, the Environmental Protection Agency, and the U.K. Mass Spectrometry Data Centre. Included for each spectrum is a "Quality Index," the compound name (as used by the Chemical Abstracts Service - CAS), the molecular weight, the molecular formula and the CAS Registry Number. This data base has already received heavy use. The computerized data base is available as a magnetic tape and can be ordered from the Office of Standard Reference Data. A published version of this data base is also available for purchase from OSRD. It is also a component of the Chemical Information System (CIS) an interagency, online data service accessible over both TYMNET and TELENET. This system has many different data bases, including mass spectra, thermodynamic properties, crystallographic parameters, x-ray powder diffraction, NMR spectra, etc. CIS is a well-established system with over 500 subscribers.

A second computerized database available from the Office of Standard Reference Data is the NBS Chemical Thermodynamic Database. This database contains the recommended values for selected thermodynamic properties for approximately 15,000 inorganic substances. These properties include the enthalpy of formation at 0° K and, at 298.15° K and 1 atm (or 1 bar), the enthalpy of formation from the elements in their standard states, the Gibbs energy of formation from the elements in their standard states, the entropy, the enthalpy content, and the heat capacity at constant pressure. In addition to these properties, the entry for each compound includes the physical state and crystal structure. A magnetic tape of these data is available from OSRD. It is one of the newest components of the Chemical Information System; the data will also appear in printed form as a supplement to the Journal of Physical and Chemical Reference Data later this year.

The NBS Crystal Data Identification File is a third computerized standard reference data file presently available. This data base contains crystallographic information on over 60,000 different crystalline materials, and includes reduced cell parameters, reduced cell volume, space group number and symbol, calculated density, classification by chemical type, chemical formula, chemical name, and an associated literature reference. In addition to crystallographic researchers, these data will be of great utility to industrial chemists, metallurgists, and pharmaceutical users. An example of its utility is in the identification of unknown crystalline substances where one has determined the primitive or centered cell, has calculated the reduced cell, and can then match this against entries in the database. In addition to being an active component on the Chemical Information System, a magnetic tape of these data is available, and printed updates are periodically published.

A fourth database is of a somewhat different nature from the above three, which contain the numeric values stored as individual data points. The Thermophysical Properties of Hydrocarbon Mixtures Database, on the other hand, consists of a small number of coefficients plus computer software to predict thermophysical properties (density, viscosity, thermal conductivity) for arbitrary hydrocarbon mixtures. The program will currently handle mixtures formed from 60 compounds, and the results have been shown to be good to approximately eight percent. The model is valid over a wide range of pressure and temperature conditions. The method is applicable to a variety of chemical types, to thermodynamic states ranging from the dilute gas to compressed liquid, and to multicomponent systems.

Above we have given a brief description of the presently available files of computerized standard reference data. An area where major emphasis is being put, indeed one of the major goals of the National Standard Reference Data System, is aimed towards the production of machine-readable data files by the data centers and projects supported by OSRD. In addition to dissemination via magnetic tape and in an online environment, such automated data files have the capability of being photocomposed, thus reducing the publication costs. Thus the same machine-readable database could be utilized in three diverse ways of dissemination – the hard-copy published form, an edition on magnetic tape (or similar media) for internal use by a given institution, and via an online network for periodic use.

The creation of these computerized databases is expensive. As elaborated upon earlier, there is more involved than just entering handbook-type tables into a machine file. The special features of numeric scientific data require specialized software while at the same time the generation of these evaluated data requires specialized personnel. The expense involved in these efforts is considerable, which is why the coordination in the development of these machine-readable databases on a national and international basis is crucial. The Office of Standard Reference Data is playing a vital role in this coordination in order to ensure that the broad data needs of the scientific and technological communities are addressed, and is placing emphasis on its work in this direction.

REFERENCES

1 Stephen A. Rossmassler, "Modification of Dissemination Channels for Scientific Information,"Journal of Chemical Documentation, IX (1969), p. 17.
2 "National Standard Reference Data System" (unpublished) pp. 1-2.

3 _Q & A About the National Standard Reference Data System_(U.S.
  Department of Commerce, National Bureau of Standards).

4 Stephen A. Rossmassler, ed.,_Critical Evaluation of Data in the
  Physical Sciences - A Status Report on the National Standard
  Reference Data System_ (Washington, D.C.: U.S. Government
  Printing Office)(1972) pp.1-2.

5 D. R. Lide, Jr., "Critical Data for Critical Needs,"_Science_,
  Vol. 212 (1981) 1343.

S.R. Heller and R. Potenzone, Jr. (Editors), *Computer Applications in Chemistry*
© 1983 Elsevier Science Publishers B.V., Amsterdam — Printed in The Netherlands

THE RANDOM-GRAPH LIKE STATE OF MATTER

JOHN W. KENNEDY

Research Institute, Advanced Medical Products, The Master's Lodge,
Dedham, Essex, ENGLAND, and the Department of Mathematics, Pace
University, New York, NY 10038, U.S.A.

INTRODUCTION

During the last few years GRAPH THEORY has clearly emerged as
a useful collection of mathematical models and tools with which to
mimic and explore the properties of interesting physical systems
that arise in a variety of disciplines. In essence, graph theory
is the simplest topological structure after that of the isolated
point (this latter structure having already found wide application
in such early chemical models as the ideal gas and the infinitely
dilute solution). The structure of graph theory is that of POINTS
and POINT PAIRS which can be conveniently visualised as LINES
connecting some pairs of the points. APPLICATIONS of this
topological structure to model physical systems require only that
the points be identified with physical OBJECTS and the lines with
RELATIONSHIPS betwen pairs of these objects (ref. 1). The power of
graph theory then lies in harnessing the vast bank of theorems
about graphs in order to gain valuable insight into the physical
system that they are being used to model. Significantly, the
complexity of the topological structure of graph theory matches
that of current physical theory in most disciplines. Furthermore,
by offering a common language, graph theory has done much to
promote the exchange of ideas among disciplines whose progress is
otherwise cloaked in specialised jargon.

Graph theory, then, is NOT A PHYSICAL THEORY in the same sense
as molecular orbital theory, etc., but rather in the sense that
mathematicians use the term. That is, it is a collection of
results and theorems that pertain to and arise as a consequence of
a well defined mathematical structure. Thus, we can speak about
"APPLICATIONS of graph theory" only in the sense that, by mapping
an observable physical system onto the graph structure (objects to

points, relationships among objects to lines) we are able to
transmute the results of graph theory into insights about the
physical system which can then be subjected to experimental
observation. Like any model, the graph theoretical structure
remains useful   only for as long as it   continues to describe our
observations and the correlations among them. The fact that graph
theory has attained its present popularity suggests that it has
much to offer in modelling physical systems to within the present
limits of our powers of observation.

As was noted at a previous meeting in Rzeszow   (refs. 1, 2)
chemical "applications" of graph theory fall into two general
headings. First, by identifying ATOMS (or sometimes small groups
of atoms) with POINTS and CHEMICAL BONDS with LINES, we map most
(if not all) of what is essential in the structure of molecules
into an area of graph theory that deals with theorems about
specific graphs (molecules) or families of graphs (or molecules).
Secondly, by identifying MOLECULES with POINTS and INTERMOLECULAR
INTERACTIONS with LINES we are enabled to use what is known about
RANDOM GRAPHS to study the statistical behaviour of large
ensembles of interacting molecules.

Examples of the first mode of employing graph theory were
cited in (ref. 1) with particular emphasis on understanding the
molecular structure and physical properties of molecules. The
numerous applications of this type are widespread in the
literature. To attempt to survey them all and at the same time to
do any of them justice must require more space than is available
here (though see references in ref. 1) and several of the poster
sessions of this meeting). For this talk, I propose to concentrate
on the second mode of applying graph theory to chemistry and
describe something of what has been found useful in random graph
theory. I shall indicate recent progress and some interesting
facets where modern computational techniques could effectively be
applied to the solution of random graph problems which, in turn,
would answer the important chemical questions that have given rise
to these problems.

Graph theoretical terms used here whose meaning is not obvious
are explained. However, those less familiar with graph theory may
benefit from consulting the excellent graph theoretical dictionary
compiled for physical scientists by Essam and Fisher (ref. 3).

RANDOM GRAPHS AND CHEMISTRY

In (ref. 2) remarks were made concerning the use of random graphs in chemistry and which bear repetition here. Chemistry often has to deal with the statistical behaviour of large ensembles of molecules or molecular fragments which interact with each other to form "clusters" held together by chemical bonds (for example in the formation of polymers), weak forces (such as interactions in fluids), or forces of intermediate strength (for example, hydrogen bonds). Our need is to be able to compute for these systems statistical quantities like the size and numbers of clusters (as a function of the physical conditions imposed on the system); statistical averages (often experimentally observable properties of the system) of the properties of individual clusters taken over the whole ensemble; and abrupt changes in these statistical quantities that accompany phase and other sharp transitions which the system might undergo and which themselves offer further, highly sensitive, experimental probes with which to enhance our understanding of molecular behaviour.

In all of the literature of this kind there pervades notions of some kind of "random structure" implicit in, but not always precisely defined by, chemical descriptions of the system. RANDOM GRAPHS offer a broad class of MATHEMATICAL models in which the imperfect notions of randomness can be precisely stated in a language designed for the task. A model, well defined in this way, can then be addressd by well developed mathematical tools that facilitate the extraction of statistical parameters. Even the task of refining random graph models, through the familiar process of inductive reasoning from a confrontation of deduced facts about the model with experimental evidence on the system, is made easier by the availability of a language suited to the task of stating the refinements needed.

Such applications of random graphs to chemistry have had success already in problems such as:

a) Molecular weight distribution of randomly branched polymers (refs. 4-8) and other properties including gelation and the post gelation rubber elasticity of random networks (refs. 9, 10). Indeed, even the more difficult problem of obtaining statistics for finite systems of this kind has recently yielded useful results (refs. 11, 12).

b) Solution thermodynamics (refs. 13, 14) and phase behaviour (ref. 15) including the recent interest in so called "lattice gas" models (refs. 16, 17) which have much importance in interpreting coal/water phase behaviour on which methods for extracting coal from deep sources depends (ref. 18).

c) The structure of the liquid state of molecular systems which, although apparently simple, have until recently (refs. 2, 19 - 22) defied adequate mathematical description. Among liquids, certainly, water is the most important and on it much has been written (refs. 23 - 29) which clearly indicates, but does not exploit, the "random graph-like" character of the structure of this system. A single, well defined random graph model for liquid water (refs. 19, 21 ,22) has tremendous implications for biological chemistry and much of what I have to say about random graphs has been motivated by work aimed in this direction.

d) The coagulation of aerosols (ref. 30) is a topic to which the ideas of random graphs are now being applied. The importance lies in the special context of predicting and controlling liquid sodium aerosols which comprise the major hazzard of nuclear reactors when leakage of their liquid sodium coolant may occur to give considerable problems down-wind of the reactor (ref. 31).

While a catalogue of chemical applications of random graph theory might be of some interest, it is less appropriate for this conference than a discussion of random graph models themselves, recent progress in computing with them, and some of the chemically interesting mathematical problems that remain. With an understanding of random graphs, participants at this meeting will be able to go forth and find many applications of their own for these ideas.

My own long-standing interest in the topic (refs. 1, 19, 20, 32, 33) has recently received further stimulus from three sources:

1) Random graphs and random graph-like models have tended to
   be studied quite separately and under a variety of names
   throughout the chemical, physical and mathematical
   literatures. There has been progress in identifying more
   clearly the various classes of random graphs and in
   clarifying the relationships and differences between them
   (see e.g. refs. 19, 20).

2) Hard graph theoretical problems are sometimes posed as
    result of attempting to apply random graphs to physical
    systems. However, there has been progress in obtaining the
    answers to some of these problems with surprising results
    (refs. 1, 19) raising further searching questions that
    strike even at our thinking about the character and
    contribution made by embedding spaces to molecular theories
    (ref. 34).

3) It has become increasingly clear that random graph models
    offer a coherent framework for understanding important
    physical systems in which dramatic and abrupt transitions
    can occur and which are often essential to the correct
    functioning of life processes (see e.g. (refs. 1, 20, 35,
    36).

Perhaps two examples will suffice to illustrate our interest
in random graphs and the elegant simplicity that underlies their
application to physical systems.

A) POLYMER GELATION:- some chemical systems initially consist
of a collection of relatively small molecules which, when they
are subjected to appropriate physical conditions, form
additional chemical bonds that link these molecules together.
This is the process of POLYMERISATION. In many cases, if it is
allowed to continue, a point is reached, called the GEL POINT,
at which a 'giant' (or gel) molecule appears and the system
rather suddenly changes from a mobile liquid to an elastic
solid. The process is familiar to anyone who has made a jelly
or boiled an egg!
    The importance of this GELATION transition to life
processes was noted by Gordon and Ross-Murphy (ref. 36). Most

of life exists in the narrow 'window' spanning the transition.
A few chemical bonds more or less would leave us too rigid or
too fluid to move comfortably.

B) THE STRUCTURE OF LIQUIDS LIKE WATER:- most substances exist
in at least three physical phases; vapour, liquid, solid. The
phase depends on the temperature (and pressure) of the system
and as this is smoothly lowered the properties of the system,
which for the most part alter gradually in response, undergo
abrupt   and   dramatic transitions when   temperatures are
encountered at which a phase change occurs (condensation
point, freezing point). The phenomenon results from molecules
interacting with each other by hydrogen bonds, Van der Waals,
and other weak physicochemical forces. Initially, in the
vapour phase, such interactions give rise to small clusters of
molecules the average order of which increases smoothly with
diminishing temperature. At the condensation temperature,
however, a 'giant' cluster (the liquid) appears suddenly and
'captures' almost all of the other molecules. As the liquid is
further cooled, interactions between molecules in the giant
cluster become more numerous causing it to become increasingly
more complex in its 'organisation' until, at the freezing
point, it dramatically and suddenly adopts a long range
ordered structure (the crystalline solid). While the vapour
and   the   crystalline solid states   are relatively well
understood, the 'random structure' of the liquid state has
presented many difficulties (ref. 37).

      Among all liquids, water occupies a uniquely interesting
place. It is the most abundant material in the Earth's crust
and comprises some 73% of the human body (80% of the brain!!).
Clearly, most of life's processes take place in liquid water.
Thus to understand these processes, we need at least to
understand how to describe the structure in terms that explain
the origin of the very unusual phenomena exhibited by liquid
water (refs. 24, 38). Subsequently, of course, we must also
assess the effects on these features resulting from the
presence of other biologically important molecules.

The process of expressing such physical systems in random
graph terms is, in a general sense, straightforward (OBJECTS to

POINTS), RELATIONSHIPS to LINES!). The principle difficulty lies in determining the class of random graphs (expressed as restrictions on the set of graphs which is to be permitted to form by a random process) so as to most appropriately describe the physical system and its chemical restrictions. In our study of random graphs we have recognised three principle and basic random graph classes as having useful applications for the statistical mechanics of molecular and other systems (refs. 2, 19, 20, 33). These classes are (see also RANDOM GRAPH CLASSES on page 158):

   i) Erdos-Renyi random graphs

  ii) Random f-graphs

 iii) Random lattice-graphs.

Applications have been discussed elsewhere in more detail (see especially (ref. 20 and references therein) and further work using recent results to answer interesting physical problems in progress (refs. 21, 22).

Because of their significance to understanding the behaviour of physical systems, two properties of all of these classes of random graphs are especially important (refs. 20, 39):

a) the existance of threshold values for the abrupt appearance of giant k-connected subgraphs (k - 1,2,...),

and

b) the distributions of cycles of length j (j-cycles).

For two of these classes these problems are now solved (refs. 19, 20, 40). For random lattice-graphs the problems remain only partly solved though they are closely related to work that has appeared, widespread and largely unconnected in the literature.

Our primary purpose here is to describe what is known about these topics and their interest for physical systems. In discussing the problems that remain to be solved for random lattice-graphs we shall attempt to relate these to chemical, physical and mathematical problems to which they are similar. By combining these various efforts and with the clever use of new computational techniques, it is to be hoped that the problems will not remain unsolved for too much longer.

RANDOM GRAPH CLASSES

RANDOM GRAPH for our purposes here refers to a graph $G_{n\ell}$ on n (labelled) points and ℓ lines picked randomly (i.e. with equal probability) from among the set $R(n, \ell : \Omega)$ of all graphs that can be constructed on n points with ℓ lines and subject to restrictions that define a particular CLASS of random graph.

In an alternative model, which we also use, our random graph is constructed from an empty graph (i.e. graph with no lines) on n points and, subject to the restrictions Ω, each possible line is added with independent probability p . To relate the two probabilistic models, we shall require that the probability p be suitably defined so that ℓ is the EXPECTED number of lines in the random graph $G_{n,\ell}$.

RANDOM GRAPH THEORY concerns itself with the STATISTICAL PROPERTIES of random graphs such as the expectation (A_G) or average value for some property A to be exhibited by $G_{n,\ell}$ properly averaged over the set $R(n, \ell : \Omega)$; or with the probability $P(A)$ that $G_{n,\ell}$ possess the property A .

In studying the EVOLUTION of a random graph, we are interested in how (A_G) (or $P(A_G)$) changes when the number of lines (or expected number of lines) ℓ in the random graph is increased.

In the work here we shall be interested only in ASYMPTOTIC results for (A_G) and $P(A_G)$ in the limit that the graph has large order (i.e. as $n \to \infty$).This asymptotic limit corresponds precisely to the so called "thermodynamic limit" in physical chemistry (ref. 41). Furthermore, in this limit the distinction between the two probabilistic models for random graphs vanishes for most purposes (and certainly all results here) since the number of lines in the second model has become sharply distributed about its expected value. Because there is no essential difference between the two models we shall feel free to use whichever is more convenient to obtain results and state all of these in terms of $G_{n,\ell}$ picked randomly from the set $R(n, \ell : \Omega)$. For further discussion on the equivalence of results see (ref. 42) and for some recent chemical work on the statistics of finite systems see (refs. 11, 12).

As in (ref. 20), our interest here concerns mainly the range of evolution in which the SIZE (i.e. number of lines) ℓ of the random graph is asymptotically proportional to n, that is:

(1) $\ell \sim c\, n$

for some constant c > 0. In this range, clearly, evolution of the random graph is reflected by increasing the value of the constant c.

The restrictions Ω determine the class of random graphs. Elsewhere (ref. 20) we have given a detailed description of the classes that interest us. We repeat the main points here. The restrictions are easily stated:

   i) (Erdos-Renyi) Random graphs, R_∞ : are SIMPLE GRAPHS (that is, they have no self-loops and no multiple lines).

   ii) Random f-graphs, R_f : are simple graphs with no point of DEGREE (or valency) greater than f.

   iii) Random lattice-graphs, $R_{L(f)}$: are simple graphs which are embeddable in some specified underlying lattice-graph L(f) of degree f and which consequently also have no point of degree greater than f), (see RANDOM LATTICE-GRAPHS on page 164 for further details).

It can be seen that these represent increasingly severe constraints on the type of graph that may be constructed by the random process.

The first two of these classes are well known in mathematical and chemical literature, respectively, (see remarks in refs. 2, 19, 20, 33). The first is, in a sense, a limiting case of the second when maximum degree restrictions are lifted (i.e. when f→∞) (ref. 33). Under this operation, any result for R_f can be translated into a result for R_∞ , though not conversely. A great deal is known about R_∞ (ref. 43), rather less is known about R_f (see refs. 2, 20 and references therein) and very few rigorous results are available for $R_{L(f)}$. In the next three sections we describe briefly some of the highlights of R_f and give results for k-connectivity and cycle distributions. We postpone until RANDOM LATTICE-GRAPHS on page 164 further discussion about random lattice-graphs.

The symbol ~ means "asymptotically equal to"; that is

$$a(n) \simeq b(n) \Rightarrow \lim_{n \to \infty} (a/b) = 1$$

RANDOM GRAPHS, RANDOM f-GRAPHS

Since the pioneer work of Erdos and Renyi (refs. 44, 45) the class of random graphs R_∞ has been well studied by mathematicians (see e.g. refs. 42, 43). Comprising simple graphs, R_∞ evolves from the empty graph on n points to a UNIQUE limit of evolution, the complete graph K_n with $\ell = \binom{n}{2}$ lines. During its evolution it passes through five distinct statistical phases (ref. 44).

In the range that interests us, see equation (1); for c <1/2, R_∞ consists of small COMPONENTS (maximal connected subgraphs) almost all of which are TREES (graphs devoid of cycles) and none of which contain more than one cycle. At c = 1/2 , R_∞ undergoes a dramatic transition: The probability that R_∞ contains a 'giant' component jumps from zero to unity. Thereafter, that is for c > 1/2, the giant component increases both in order (number of points) and in complexity while the remaining small tree components are 'captured' by it.

In general terms the evolution of R_∞ resembles physical and chemical processes (ref. 20), but, for the most part, chemical applications of random graph theory require that degree restrictions be imposed. Chemists had, under other guises, independently studied random f-graphs R_f (refs. 2, 19, 20). These, like R_∞ , also evolve from an empty graph on n points. However, the limit of evolution is now not unique but consists of one of the f-regular graphs (i.e. graphs in which every point has degree f) selected randomly from the set of all f-regular graphs on n points. Furthermore, this limit of evolution has $\ell \sim fn/2$ and so lies within the range of interest to us (equation 1). In general terms, the evolution of R_f is similar to that of R_∞ . The abrupt transition leading to appearance of the giant component is now delayed until c = f/(2(f-1), a consequence which can be observed directly in polymer systems and (at least) indirectly in the condensation of vapours into liquids (refs. 2, 19, 20).

For the remainder of this paper it will be convenient to deal with R_∞ and R_f together, treating R_∞ as a limiting case of R_f when $f \to \infty$.

k-CONNECTED TRANSITIONS IN RANDOM f-GRAPHS

A graph is said to be k-CONNECTED if it requires that a set of at least k lines have to be removed from it to disconnect the graph into more than one component. As was noted above, when a random f-graph evolves it exhibits an abrupt transition (or threshold) related to the sudden appearance of a giant component or, equivalently, a GIANT 1-CONNECTED SUBGRAPH. The phenomenom is closely analogous to transitions that occur in random graph-like physical systems such as the gelation of polymers or the condensation of vapours into liquids (refs. 2, 19, 20), to the sudden onset of magnetism in ferromagnetic materials (ref. 46), and to the point of no return in the criticallity of a thermonuclear reactor.

Kennedy suggested (ref. 39) and with Erdos later answered (ref. 40) an interesting generalisation of this phenomenom concerning thresholds for the abrupt appearance of giant k-connected subgraphs (k =1,2,...,f) in random f-graphs. That is, in R_f with ℓ ~ cn (equation 1), for each k = 1,2,...,f, there exists a critical value c_k such that:

(2) Probability that R_f $\sim$ $\begin{cases} 0 & \text{if} \quad c < c_k \\ 1 & \text{if} \quad c > c_k \end{cases}$
     contains a giant $n \to \infty$
     k-connected subgraph

For k = 1 we have already encountered this as the appearance of a giant component (see RANDOM GRAPHS, RANDOM f-GRAPHS on page 160) and $c_1 = f/(2(f-1))$. More generally (ref. 40):

$$(3) \quad \frac{f}{2(f-1)} = c_1 = c_2 < c_3 < \dots c_{f-1} < c_f = \frac{f}{2(f-1)^{1-(f-1)}}$$

where c_k is the solution (between 0 and f/2) to:

$$(4) \quad \sum_{j=k-1}^{f-1} j \binom{f-1}{j} (1 - 2c_k/f)^{f-1-j} (2c_k/f)^j = 1$$

For R_∞ we can show directly (ref. 40) or obtain from equations (3) and (4) by letting $f \to \infty$ that :

(5) $1/2 = c_1 = c_2 (c_3 (c_4 < \dots$

where c_k is the solution to

(6) $2c_k - e^{-2c_k} \sum\limits_{j=0}^{k-2} j(2c_k)^j/j! = 1.$

We reproduce some values for c_k (ref. 40) in TABLE 1.

TABLE I. Threshold values c_k of c (Equation 1) for the appearance of a giant k-connected subgraph in R_f (and R_∞).

f \ k	1,2	3	4	5	6	7	8	9	10	11	12
3	0.750	1.061									
4	0.667	0.933	1.387								
5	0.625	0.867	1.250	1.768							
6	0.600	0.828	1.175	1.615	2.175						
7	0.584	0.802	1.128	1.528	2.004	2.597					
8	0.572	0.783	1.095	1.471	1.905	2.410	3.029				
9	0.652	0.770	1.072	1.430	1.838	2.299	2.828	3.470			
10	0.556	0.759	1.053	1.400	1.789	2.223	2.706	3.255	3.917		
11	0.550	0.750	1.039	1.376	1.752	2.166	2.621	3.123	3.690	4.369	
12	0.546	0.743	1.024	1.357	1.723	2.123	2.557	3.029	3.548	4.130	4.82

The coincidence of c_1 and c_2, while unexpected, has an obvious explanation in terms of the creation of cycles of infinite order (giant 2-connected subgraphs) and we remark on this further in the next section.

The higher thresholds, like c_1, appear to have analogues with transitions in physical systems (ref. 20). Thus analogy with the transitions in R_4 strongly suggests that the water/ice transition is associated with c_4 (refs. 21, 22) and at temperatures close to that where $c = c_3$, liquid water does exhibit subtle anomalies (refs. 24, 38) which have yet to be adequately explained (refs. 21, 22). A detailed treatment of these phenomena will, however, have to await the solution of the k-connected threshold problem in random lattice-graphs (see k-CONNECTED TRANSITIONS on page 164).

CYCLE DISTRIBUTIONS IN RANDOM f-GRAPHS

The evolution of random graphs beyond the threshold c_1 is of interest in relation to their use as models for the behaviour of physical systems (refs. 2, 19, 20). For $c > c_1$, R_f consists of a single giant component and a distribution of trees of various orders. As c increases (above c_1) the giant component (i) grows in order, by 'capturing' tree components with the largest trees most likely to be 'captured' first; and (ii) gains an increasing number of cycles. At any c both the order of the giant component and the distribution of orders among the remaining tree components in R_f can be determined (see e.g. refs. 2, 19). What is of interest, especially in relation to interpreting the behaviour of liquids like water is the distribution of cycles of various lengths in the giant component (refs. 39, 47). Since the giant component contains ALL the cycles in R_f this is equivalent to asking for the expected number of cycles of length j (j-cycles) in the random graph.

Consider a random f-graph R_f on n points and with $\ell \sim cn$ lines, then (ref. 19): The asymptotic probability that R_F contains exactly t j-cycles is the coefficient of θ^t in the probability generating function:

$$(7) \qquad C_j(\theta) \sim \sum_t \frac{\lambda_j^t e^{-\lambda_j}}{t!} \theta^j = e^{\lambda_j(\theta-1)}$$

where

$$(8) \qquad \lambda_j \sim \frac{1}{2j} \left(\frac{2(f-1)}{f} c\right)^j$$

is the expected number of j-cycles in R.

Similarly, for R_∞ in the domain of evolution for which equation (1) applies, let $f \to \infty$ in the above and the expected number of j-cycles is seen to be:

$$(9) \qquad \lambda_j \sim \frac{(2c)^j}{2j}$$

We note from equation (8) that, when $c < c_1$ (= f/2(f-1)), the expected number of j-cycles is a decreasing function of j so that long cycles are almost surely absent from R_f (or R_∞). However, for $c < c_1$, λ_j is an increasing function of j and cycles of all lengths appear. Apparently c_1 is also a threshold for the appearance of large (giant) cycles and so coincides with c_2 the

threshold for a giant 2-connected subgraph; a feature which was
demonstrated independently in the preceeding section.

The final point of interest concerns the absence of n from the
asymptotic expressions (ref. 7, 8). As has been remarked (refs. 2,
3, 4) this generates some surprises since it does not accord with
experimental observations in many physical systems for which
random f-graphs had found wide acceptance as models (refs. 2, 19).
For example, applications to theories of rubber elasticity and
even more so in the theory of liquids as they approach the
crystalline transition. It was for this reason that we found the
need to use random lattice-graphs for these applications and
thereby ensure that as the limit of evolution is reached the
number of j-cycles approaches what is expected on physical
grounds. We turn to cycle distributions in random lattice-graphs
in CYCLE DISTRIBUTIONS IN $R_{L(f)}$ on page 168.

RANDOM LATTICE-GRAPHS

We have noted (paragraph 2) that a random lattice-graph $R_{L(f)}$
is a graph drawn randomly from the class of graphs on n point with
lines and which are embeddable in an underlying lattice-graph L(f)
of degree f . For our purposes a LATTICE-GRAPH is an infinite
order, connected, point-transitive, line-transitive, regular graph
but see (refs. 3, 34) for further discussion. A graph H is
EMBEDDABLE (ref. 34) in L(f) iff a mapping of H into L(f) can be
be found that:

(i) sends distinct points of H onto distinct points in L(f)

and

(ii) sends adjacent points of H onto adjacent points in L(f).

The evolution of $R_{L(f)}$ again starts with the empty graph on n
points and proceeds to its limit which, like R_∞ but unlike R_f ,
is a UNIQUE graph. The unique limit, like R_f , has $\lambda \sim fn/2$ lines.
It has a regular structure (that of the underlying lattice-graph
L(f). It is this limit of the evolution process that makes $R_{L(f)}$ a
better model for liquids at temperatures extending down to that of
the transition at which they freeze to form crystalline solids. By
choosing the appropriate lattice-graph L(f) the correct limiting

cycle distribution is guaranteed; a fact that has already been used to explain the maximum density shown by liquid water at $4^{\circ}C$(refs. 2, 20, 21) and hence the reason why icebergs float (ref. 47). Furthermore, the explanation is made in terms of a model which is consistent with other observable features of water and of similar liquids (refs. 21, 22).

Criteria for embeddability of f-trees (i.e. trees with maximum degree f) in L(f) were studied by Kennedy & Quintas (ref. 34) from the viewpoint of suitability of embedding spaces for molecular graphs in chemical theories. Schurger (ref. 48) introduced and studied a class of random graphs closely related to our random lattice-graphs using the SQUARE lattice-graph and obtained useful statistical results for this case. Random graphs of this type have also been studied in some detail under the topic of percolation theory (refs. 49 - 55) but, in contrast to the other two classes of random graphs discussed above, few exact results are available relating to abrupt transitions and to cycle distributions in $R_{L(f)}$.

Nevertheless, some partial results are available although the effort to glean these results from the literature is hampered by the fact that they are widespread and couched in a variety of disguises and by many mathematical techniques such as percolation theory (refs. 49 - 55), lattice-walks and paths theory (refs. 56 - 61), polyominoe theory (refs. 62 - 67), clustering theory (refs. 50, 68, 69), to name but a few. In the following sections we shall try to summarise what is known relating to k-connected transitions and cycle distributions in $R_{L(f)}$.

k-CONNECTED TRANSITIONS in $R_{L(f)}$

The evolution of $R_{L(f)}$ shows a general resemblance to that of R_f. The point degree distribution probability generating function for $R_{L(f)}$ is identical to that for R_f throughout the evolution, namely: The expected probability for a point selected randomly from $R_{L(f)}$ (or R_f) to have degree x is just the coefficient of θ^x in the pgf (see ref. 19):

(10) $F(\theta) = (1 - \dfrac{2c}{f} + \dfrac{2c}{f} \theta)^f$.

For small enough ℓ , $R_{L(f)}$ is indistinguishable from R_f since, in the very early stages of evolution, almost all components are

f-trees which can be embedded in L(f)   (cf. ref. 34). As evolution
proceeds, however, the two   models   diverge. Although $R_{L(f)}$ does
undergo abrupt transitions associated with the appearance of giant
k-connected subgraphs (ref. 40)   much   less is   known about   the
statistical properties and thresholds for these transitions.

The 1-connected transition   (formation of a   giant   component)
occurs at a value   for the constant c_1 (cf. equation ref. 2) that,
not suprisingly, depends on the underlying lattice-graph L(f). In
contrast to R_f (or R_∞) it is not the case that formation of
cycles can be neglected on approaching the transition in $R_{L(f)}$.

The   transition   value   of   the constant, c_1 , in $R_{L(f)}$ is
essentially related   to the so   called   "critical probability   for
percolation" on L(f) (refs. 50,52). The most extensive collection
of critical bond   probabilities $p_c^{(B)}$ for   various   lattice-graphs
appears to be   that of Essam (ref.   50).   If p is   the independent
probability   for   a line in $R_{L(f)}$ then the number of lines   is
related to it thus:

(11) $\ell \sim pfn/2 \sim cn$.

Consequently, if $p_c^{(B)}$ is the critical bond probability (ref. 50) for
percolation on L(f) , then the critical value of our constant c_1
is:

(12) $c_1 = f\, p_c^{(B)}/2$.

We   give these critical values in TABLE 2   for some lattice-graphs
of   physical   interest,   in   comparison   with   values   for   the
corresponding random f-graphs.

For $R_{L(f)}$, although   the higher transitions do occur, their
location remains an important unsolved problem. Of course, for all
k it seems intuitive that the transition in $R_{L(f)}$ should occur at
a value for the constant c_k that is higher than the corresponding
value   known (see k-CONNECTED TRANSITIONS   IN   RANDOM f-GRAPHS   on
page 161) for R_f . This is for two reasons:

TABLE 2.   Critical values c_1 in equation (ref. 2) for the formation of a giant 1-connected subgraph (component) in random lattice-graphs $R_{L(f)}$.

L	f	$P_c^{(B)}$ from ref. 50, p224	Critical Value for $R_{L(f)}$ text eqn. 12	Critical Value for R_f text page 161
hexagonal	3	0.6527 (exact)	0.8796	0.75
square	4	0.5000 (exact)	1.0000	0.6667
triangular	6	0.3473 (exact)	1.0419	0.6000
diamond	4	(0.388)* 0.005	(0.798)* 0.01	0.6667
cubic	6	(0.247)* 0.005	(0.741) 0.01	0.6000

* approximate values from series method ref. 50.

i) Let $d_{r,v}$ be the number of points in a graph G at a distance r from some point $v \in G$. Then the sequence $\{d_{1,v}, d_{2,v}, d_{3,v}, \ldots\}$ is called the distance degree sequence of v in G (refs. 34, 70). The additional constraints imposed on $R_{L(f)}$ as compared with R_f are of the type that limit the allowed distance degree sequences that can be found in $R_{L(f)}$ (see ref. 34). Thus these constraints represent even more stringent limitations on the possible distance degree sequences than those imposed in passing from R to R_f which limited only $d_{1,v} \leq f$. The effect is to exclude highly branched tree-like subgraphs of small diameter and as a result defer the appearance of the giant component to a later stage in the evolution of $R_{L(f)}$

ii) There is a much enhanced probability of cycle formation in $R_{L(f)}$ as compared with R_f (the very reason why random lattice-graphs were introduced). Since each cycle formed consumes a line in the random graph without increasing the order of any component, more lines are required (that is c has to be increased) to compensate for the lines 'wasted' in cycle formation, thereby delaying the transition leading (for example) to a giant component.

The two effects described above are highly interrelated and thus further progress on the problem of k-connected transitions in random lattice-graphs, though important from a physical point of view, will very likely have to await progress on the problems of cycle distributions, the topic to which we now turn.

CYCLE DISTRIBUTIONS IN $R_{L(f)}$

The distribution of cycle lengths in random lattice-graphs emerges as the most urgent problem that now needs to be solved for the meaningful application of random graph theory to physical systems.

Consider a random lattice-graph $R_{L(f)}$, with underlying lattice-graph L(f) of degree f and GIRTH (length of shortest cycle) g. Let $Z_j(L)$ be the number of distinct j-cycles (cycles of length j) in L(f) . Then, for $R_{L(f)}$ the expected number of j-cycles $\langle \lambda_j \rangle$ is:

(13) $\langle \lambda_j \rangle \sim (2c/f)^j \, Z_j(L)$

so that the statistical problem in random lattice-graphs now becomes the COMBINATORIAL problem of finding the sequence of numbers $\{Z_j : j =1,2,...\}$ for L(f) , a problem which is still unsolved in general (see for example (refs. 19, 48, 50, 52, 56, 68).

For all lattice-graphs L(f) it is clear that:

(14) $Z_j(L) \sim y_j \, n$

where y_j is a constant depending on j and L(f) . This at least renders explicit the fact that the expected number of j-cycles in

$R_{L(f)}$ is asymptotically proportional to n (the number of points) in satisfacsfaction of our needs for suitable physical models (refs 2, 19 see also RANDOM LATTICE-GRAPHS on page 164).

For most lattice-graphs it is again obvious that cycles of some lengths are absent; that is, $y_j = 0$ for some j as is illustrated in TABLE 3.

TABLE 3. Values of j for which there are no j-cycles (i.e. $j_j = 0$) in lattice-graphcs L(f) of degree f and girth g.

L(f)	f	g	$y_j = 0$ for
square	4	4	
			j < 4 and all odd j
cubic	6	4	
hexagonal	3	6	
			j < 6, j=8 and all odd j
diamond	4	6	
triangular	6	3	j < 3

Some other exact results are available. Thus, the first few values of y_j have been obtained for some lattice-graphs by inspection or by exact computer counting methods (refs. 57, 58). We illustrate this in TABLE 4 for the square lattice-graphs.

The series for y_j of the square lattice-graph extends thus:

(15)   1, 2, 7, 28, 124, 588, 2938, 15268, ...

(see ref. 71) and it, like other exact enumeration results of this kind, have appeared mainly in the physical chemical literature under the topic "lattice walk enumeration", from which the best results known to us are:

square lattice-graph, up to j = 18 (ref. 56)
cubic lattice-graph, up to j = 14 (ref. 56)
diamond lattice-graph, up to about j =20 (refs. 68, 72).

A wealth of closely related work also exists in the literature
under the heading of "lattice animals" (ref. 68) or "polyominoes"
(refs. 62-67). Limiting ourselves to the square lattice only for
clarity, a square-lattice animal (or polyominoe) is constructed by

TABLE 4. Number of distinct j-cycles in the square lattice graph.

j	Type of Cycle and Multiplicity	y_j

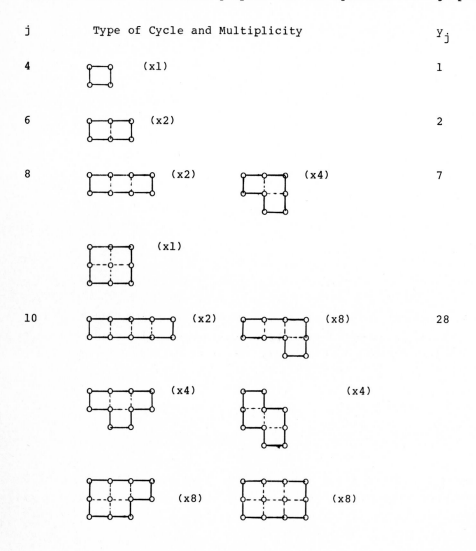

j		y_j
4	(x1)	1
6	(x2)	2
8	(x2) (x4) (x1)	7
10	(x2) (x8) (x4) (x4) (x8) (x8)	28

adjoining pairs of unit squares at their edges. An animal of order
n (or n-ominoe) results from connecting together n such squares in
this way and the animal can be said to have "area" n . The

"perimeter" of an animal is the number of its peripheral edges
(i.e. edges NOT shared with adjacent squares). TABLE 4 illustrates
all the distinct "shapes" of (simply connected) square animals
with perimeters 4, 6, 8 and 10 respectively.

With this background, the problem of finding the   sequence $\{y_j\}$
for a lattice-graph can now be transformed into two (possibly)
simpler problems.

1) What is the multiplicity m of an animal (polyominoe) of
fixed shape S in the underlying lattice-graph L(f) ; that is,
how many distinct orientations are possible for the shape S in
L(f) (see TABLE 4 for an illustration) ?

2) What is the number of animal "shapes"; that is, how many
simply connected animals of fixed PERIMETER does L(f) induce ?

The first of these two problems is relatively straightforward.
For any planar-embeddable lattice-graph L(f) , we have:

(16)   m = 2f/s

where s is the order of the symmetry group of the shape S .
Furthermore, a similar result can be assumed to obtain for non
planar-embeddable lattice-graphs.

The sequence for the number of square animal shapes of fixed
perimeter (= 1,2,3,...) is (cf. TABLE 4 and equation 15):

(17)   1, 1, 3, 6, 25, ...

but regretfully, almost all of the work on animals (polyominoes)
has concerned the numbers of animals with fixed AREA rather than
perimeter and thus it remains a potentially productive avenue for
approaching the problem of cycles in random lattice-graphs.

Exact enumeration results for the sequence $\{y_j\}$ have already
proved useful in some physical problems: For example, the y_6
value for the diamond lattice-graph was used in applying random
lattice-graphs to explain the maximum density of water at $4\,°C$
(ref. 2).

For other important physical problems, especially of the type that involve critical phenomena, what is required are values of y for large j . Obviously, any direct counting of cycles rapidly becomes prohibitively costly in terms of computer time as j increases.

For large cycles, then, our best hope at present lies in obtaining approximate formulae for y_j for which a number of methods suggest themselves:

1) Extrapolate to large j from the exact cycle counts that are available see e.g. (ref. 56).

2) Make use of statistical estimation techniques such as Monte Carlo to obtain values for large j.

3) Obtain estimates of the number a_j of animals with perimeter j and then, noting that for large j they are 'almost all' asymmetric (i.e.s=1), apply equation (16) and obtain $y_j \sim 2fa$.

Because of the insight which these results will give to the behaviour of physical systems, they are being actively persued and it will be a pleasure to be able to report progress on some future occasion.

CLOSING REMARKS

I hope to have been able to give some insight into the wide applicability of graph theory and random graph theory in particular, to chemical and physical systems, as well as its utility in understanding them. Indeed, I hope more to have given some understanding of random graphs themselves so that they may find new uses as a result of that mysterious fermentation process that goes on in all our minds. For me, as exciting as the things that random graph theory CAN say about physical systems, are the problems that these systems have suggested to graph theory. The pursuit for the sometimes elusive answers to these problems is itself fascinating (and at times frustrating) but that pleasure is surely nothing in comparison with the rewards that their solution will bring in terms of squeezing out a bit more understanding of the real world and some of life's most essential processes. This, of course, is one of the keystones of meaning for all academic pursuits.

Eventually, the RANDOM-GRAPH-LIKE STATE OF MATTER will, like its predecessors, have to be abandoned in favour of a more intricate theory: But that day is not yet. Unlike 'polywater' which Hildebrand found "a bit hard to swallow", random-lattice-graph-like water still offers an interesting blend of ideas that glide smoothly across the palate.

ACKNOWLEDGEMENTS

The author thanks Dr. Steve Heller and the organisers of the 6th International Conference on Computers in Chemical Research and Education for the invitation to present these thoughts.

Thanks are also due to the Directors of Advanced Medical Products for their vision in supporting the work of the Research Institute; and to the Pace University Mathematics Department for their encouragement of this work and flexibility in logistic arrangements for my presently spasmodic visits to them.

Finally, expecial thanks are due to my colleagues Professors Gary Bloom (City college, CUNY) and Louis Quintas (Pace University) and Mgr. Malgorzata Mandziuk-Kennedy (Advanced Medical Products) whose involvement in the work presented here made most of it possible.

The work itself is dedicated to OLIVER ANDRZEJ KENNEDY on the occasion of his zeroth birthday, May 29, 1982.

REFERENCES

1 J.W. Kennedy, Small Graphs, Graph Theory and Chemistry, in DATA
   PROCESSING IN CHEMISTRY, Rzeszow, Poland (1980). Proceedings,
   Elsevier,in press.

2 J.W. Kennedy, Statistical Mechanics and Large Random Graphs, in
   DATA PROCESSING IN CHEMISTRY, Rzeszow, Poland (1980),
   Proceedings, Elsevier, in press.

3 J.W. Essam & M.E. Fisher, Some Basic Definitions in Graph
   Theory, Reviews of Modern Physics, 42 (1970) 272-288.

4 M. Gordon & W.B. Temple, The Graph-Like State of Matter and
   Polymer Science, Ch.10, 299-332 in Chemical Applications of
   Graph Theory, A.T. Balaban (Ed.), Academic Press (1976).

5 M. Schmidt and W. Burchard; Applications of Cascade Theory to
   Calculation of Quasielastic Scattering Functions: 2. -
   Polydisperse Branched Molecules in Dilute Solution.
   Macromolecules, 11 (1978) 460-465.

6 K. Dusek, Statistika tvorby Sesitovanych Polymeru (Statistics
   of Crosslinked Polymer Formation). Edice Macro Rada R-Revue,
   Svazek R-1 (1978) - in Czech.

7 D. Durand & C.-M.Bruneau; Graph Theory and Molecular
   Distribution: 2. - Copolymerisation of A-group Polyfunctional
   monomers with B-group Polyfunctional Monomers. Macromolecules,
   12, (1979) 1216-1222.

8 W. Klonowski, Probabilistic Theory of Structural and Physical
   Characteristics of Crosslinked Polymer Systems. Rheol. Acta, 18
   (1979) 442-450.

9 A. Ziabicki and J. Walasek; Topological Structure and Physical
   Properties of Permanently Cross-Linked Systems: 1.-
   s-functional, Homogeneous, Gaussian systems. Macromolecules, 11
   (1978) 471-476.

10 K. Dusek and M. Ilavsky; Problems of Structural
   Characterisation of Polymer Networks. Polymer Engineering
   Science, 19 (1979) 246-253.

11 E. Donoghue and J.H. Gibbs, Condensation Theory for Finite,
   Closed Systems, (Department of Chemistry, Brown University,
   Providence, R102912, USA). Master of Science Dissertation
   (1980)

12 E. Donoghue, Analytic Solutions of Gelation Theory for Finite,
   Closed Systems, (Department of Mathematics, Amherst College,
   Amherst, Mass. 01002. Master of Science Dissertation (1981)

13 J.W. Kennedy, Thermodynamics of Solutions and Mixtures, Review
    for Specialist Periodical Reports, Royal Society of Chemistry,
    Macromolecular Chemistry, (1980) Vol.1, Ch. 14, 296-329.

14 P.A. Irvine and J.W. Kennedy; Efficient Computation of Phase
    Equilibria in Polydisperse Polymer Solutions Using R-Equivalent
    Delta-Function Molecular Weight Distributions, Macromolecules,
    15 (1982) 483-482.

15 D.S. McKenzie; Polymers and Scaling, Physics Reports, 27 (1976)
    35-38.

16 C. Domb, Metastability and Spinodals in the Lattice Gas Model,
    J. Phys. a, 9 (1976) 283-299.

17 L.A. Kleintjens, Effects of Chain Branching and Pressure on
    Thermodynamic Properties of Polymer Solutions, (Department of
    Chemistry, Essex University) Ph. D. Dissertation (1979).

18 L.A. Kleintjens and R. Koningsveld, Equation of State from
    Phase Equilibria, Chemical Technology Meeting, Berlin,
    Proceedings in Preparation (1980).

19 J.W. Kennedy, Icycles-I (Random Graphs, Physical Transitions,
    Polymer Gels, and the Liquid State), in Theory and Applications
    of Graphs, G. Chartrand et al (Eds.), John Wiley & Son, NY,
    (1981) 409-430.

20 G.S. Bloom, J.W. Kennedy, M.T. Mandziuk and L.V. Quintas,
    Random Graphs and the Physical World, International Conference
    on Graph Theory Dedicated to the Memory of Kazimierz
    Kuratowski, Lagow, Poland , Springer-Verlag, Proceedings in
    preparation (1981).

21 L.V. Quintas, G.S. Bloom and J.W. Kennedy, Icycles-II (Random
    Lattice-Graphs: A Model for Liquid Water). In preparation.

22 M.T. Mandziuk-Kennedy and J.W. Kennedy, Icycles-III (A Model
    for Liquids Like Water). In preparation.

23 D. Eisenberg & W. Kauzmann, The Structure and Properties of
    Water, Oxford University Press, London (1969).

24 R.A. Horne (ed.); Water and Aqueous Solutions: Structure,
    Thermodynamics and Transport Processes, Wiley-Interscience
    (1972).

25 D.L. Beveridge, M. Mezei, S. Swaminathan and S.W. Harrison,
    Monte Carlo Studies of the Structure of Liquid Water and
    Aqueous Solutions, in Computer Modelling and Matter, P.Lykos
    (ed.), American Chemical Society Sympos. Series, 86 (1978)
    191-218.

26 M. Mezei and D.L. Beveridge, Theoretical Studies of Hydrogen
   Bonding in Liquid Water and Dilute Aqueous Solutions,
   J.Chem.Phys. (1980).

27 R. Lenk, M. Bonzon and H. Geppin, Dynamically Oriented
   Biological Water as Studied by NMR, Chemical Phys. Letters, 76
   (1980) 175-177.

28 S.A. Rice, W.G. Madden, R. McGraw, M.G. Sceats and M.S.
   Bergren, On the Relationship between Low-Density Amorphous
   Solid Water and Ice Ih, J. of Glaciology, 21 (1978) 509-535.

29 S.A. Rice and M.G. Sceats, A Random Network Model for Water,
   J.Phys.Chem., 85 (1981) 1108-1119.

30 S.J. Rooker and C.N. Davies, Measurement of the Coagulation
   Rate of a High Knudsen Number Aerosol with Allowance for Wall
   Losses, J. Aerosol Sci., 10 (1979) 139-150.

31 H.M. Kottowski (Ed.), Safety Problems Related to Sodium
   Handling in Liquid Metal Fast Breeder Reactors and Large Test
   Facilities, Lectures from a course, Ispra, Italy Harwood
   Academic Publishers, Holland (1980).

32 W. Burchard, K. Kajiwara, M. Gordon, J. Kalal and J.W. Kennedy,
   Macromolecules, 6 (1973) 642-649.

33 J.W. Kennedy; Random Clumps, Graphs, and Polymer Solutions, in
   Theory and Applications of Graphs, Y.Alavi & D.Lick (Eds.),
   Proceedings Michigan (1976). Lecture Notes in Mathematics,
   #642, Springer-Verlag, Berlin 1978.

34 J.W. Kennedy and L.V. Quintas, J. Discrete Applied Math., in
   press.

35 J.E. Cohen, Threshold Phenomena in Random Structures, 7th
   International Conference From Theoretical Physics To Biology,
   Vienna (1979).

36 M. Gordon & S.B. Ross-Murphy; The Structure and Properties of
   Molecular Trees and Networks, Pure & Applied Chem., 43, (1975)
   1-26.

37 J. A. Pryde; The Liquid State, Hutchison University
   Publications, London (1966).

38 M.C.R. Symons, M.J. Blandamer and M.F. Fox, New Scientist, May
   11, (1967), 345-346.

39 J.W. Kennedy, k-Connectivity and Cycles in Random Graphs with
   Applications, Notes from New York Graph Theory Day I, (GTD
   I:2), May 1980, New York Academy of Sciences (1981).

40 P. Erdos and J.W. Kennedy, k-Connectivity in Random Graphs, Submitted 1981.

41 L.E. Reichl, in A Modern Course in Statistical Physics, University of Texas Press, Austin (1980),pp 195 and 205.

42 B. Bollobasl ch.7 in GRAPH THEORY, Springer-Verlag 1979.

43 M. Karonski, J. Graph Theory, in press.

44 P. Erdos and A. Renyi, Bull. Inst. Internat. Statist. (Tokyo) 38 (1961) 343-347.

45 P. Erdos, The Art of Counting, J. Spencer (Ed.) MIT Press, Ch. 14, pp. 559-611, 1973.

46 M. Vicentini-Missoni, Equilibrium Scaling in Fluids and Magnets, Ch.2, pp.39-77 in Phase Transitions and Critical Phenomena, Volume 2, C.Domb & M.S.Green (Eds.), Academic Press (1972).

47 J.W. Kennedy, Graph Theory Newsletter, 9 (1979) 5-6.

48 K. Schurger, Acta Mathematica Scientiarium Hungaricae, 27 (1976) 281-292.

49 H.N.V. Temperley and E.H. Lieb, Proc. Roy. Soc. Lond. A, 322 (1971) 251-280.

50 J.W. Essam, in Phase Transitions and Critical Phenomena, Vol.2, C.Domb, M.S.Green (Eds.), Academic Press (1972) 197-269.

51 J. Kurkijarvi and T.C. Padmore, J.Phys. A: Math.Gen., 8, (1975) 683-696.

52 D.J.A. Welsh, Percolation Theory and Related Topics, Science Progress, Oxford, 64 (1977) 65-83.

53 P.D. Seymour and D.J.A. Welsh, Annal. Discrete Math., 3 (1978) 227-245.

54 D.S. Gaunt, A.J. Guttmann and S.G. Whittington, J. Phys. A: Math. Gen., 12 (1979) 75-79.

55 J. Hoshen, Chemical Phys., Letters, 75 (1980) 345-349.

56 G.S. Rushbrook and J. Eve, J. Chem. Phys., 31 (1959) 1333-1334.

57 J.L. Martin, Proc. Camb. Phil. Soc., 58 (1962) 92-101.

58 M.L. Huggins and J.W. Kennedy, Polymer J. (Japan), 11 (1979) 315-322.

59 C. Domb, Adv. Chem. Phys., 15 (1969) 229-260.

60 M. Kumbar and S. Windwer, J. Chem. Phys, 50 (1969) 5257-5261.

61 N.C. Jain and W.E. Pruitt, J. Analyse Math, 24 (1970) 369-393.

62 S.W. Golomb, Polyominoes, Charles Scribner's Sons, New York (1965).

63 W.E. Philpott, Recreational Math., 10 (1977/1978) 2-14.

64 W.E. Philpott, Recreational Math., 10 (1977/1978) 98-105.

65 W.F. Lunnon, in Graph Theory and Computing, R.Read (Ed.)
    87-89.

66 W.F. Lunnon, in Graph Theory and Computing, R.Read (Ed.)
    101-108.

67 A.L. Clarke, Recreational Math., 13 (1980-81) 18-25.

68 C. Domb, T. Schneider and E. Stoll, J. Phys. A: Math.Gen., 8
    (1975) L90-L94.

69 N. Biggs, Quarterly J. Math., 29 (1978) 159-173.

70 G.S. Bloom, J.W. Kennedy and L.V. Quintas, in Theory and
    Applications of Graphs, G. Chartrand et al (Eds.), John Wiley &
    qons, NY, (1981) 95-108.

71 N.J. Sloane, Handbook of Integer Sequences, sequence #703,
    Academic Press, New York (1973) 78.

72 C. Domb, Private Communication.

S.R. Heller and R. Potenzone, Jr. (Editors), *Computer Applications in Chemistry*
© 1983 Elsevier Science Publishers B.V., Amsterdam — Printed in The Netherlands

THE FEDERAL ROLE IN THE SUPPORT OF U.S. SCIENCE AND TECHNOLOGY:
A VIEW FROM CONGRESS

CONGRESSMAN GEORGE E. BROWN, JR.
36th District, California

Thank you for inviting me to speak tonight. I am here as a replacement for Congressman Don Fuqua, Distinguished Chairman of the Science and Technology Committee on which I serve, who very much regrets that an unforeseen conflict prevents his attendance. Don has done an excellent job of steering the Committee through a difficult budgetary period. It seems particularly difficult in these times of budget constraint to feel that our work on the so-called authorizing, or policy, committees of Congress has much impact on programs. Often our policy debates are clouded and obscured in the overall budget process.

Many observers of the science scene, including the President's Science Advisor Dr. George Keyworth and Chairman Fuqua, believe that science and particularly basic research has fared well. They are being kinder than I would be under the circumstances. But they may have a different perspective than I do. As merely one member of the Science and Technology Committee, I will sometimes allow myself to vent my feelings on what is happening in science and technology on Capitol Hill, and what I think ought to be happening.

When the Reagan Administration came to town it brought with it two overriding principles:

One --THAT THIS NATION'S DEFENSE CAPABILITIES NEED BOOSTING.

Two--THAT THE WAY TO ACHIEVE ECONOMIC RECOVERY IS TO GET THE FEDERAL GOVERNMENT OUT OF THE WAY OF BUSINESS AND LET THE PRIVATE SECTOR TAKE OVER.

Whether or not one agrees with the President, there is no
doubt that his philosophy can have and is having a direct impact
on the science and technology efforts supported by the federal
government.

A major science policy decision has been taking place the last
one and one-half years which has been obscured partly by debates
over federal budget priorities and federal budget deficits. Put
starkly, the question has been "Should the federal government be
involved in a significant way in non-defense science and
technology?" Programs ranging from science education to space
exploration have been cut back dramatically to conform with the
Administration's principles, while using the pretext of budget
constraints.

Before I continue, I need to make a few disclosures. I am a
Democrat who believes that government can be part of the solution
to the problems of society. I feel particularly strongly that the
federal government's support of science can underpin our nation's
success at home and internationally in areas such as agriculture,
medicine, communications, energy and space. And although the cuts
in this year's science budget have not been as dramatic as in
other federal civilian programs, on close examination, science
programs across the board are suffering.

When one looks at long-range budget projections, the growth in
the defense budget and the reduction in the discretionary
(NON-ENTITLEMENT) parts of the civilian budget, pressure on
research and development support is evident.

The space program is a good example of this trend away from
civilian research to military research in our science programs.
While the space program may be the only federal program aside from
defense to be increased, this increase is targeted for NASA's
space shuttle program which now, according to the President's
newly announced Space Policy, will provide priority for national
security missions, although a civilian agency is supporting it
financially. Planetary science and space applications funding will
suffer substantial cuts.

Perhaps the most potent political issue in science policy this
year is the current crisis in science and engineering education.
While education has traditionally been under the jurisdiction of
the states, the Congress has in certain cases felt a need for
federal support in areas such as student aid, aid for the

handicapped, or special programs for the gifted.

The federal government clearly became interested in science education after the launching of Sputnik by the U.S.S.R. and an aggressive science education effort was launched. I think this country may be more threatened technologically now than at that time if only because our economic situation seems to have thwarted our will to take positive and effective action in the science education area.

Although the United States rates high in Nobel Laureates, it is not unreasonable to ask if the policy changes now suggested will substantially reduce the ability of American researchers to compete for the Nobel Prizes in the 1990s and following years. The groundwork for present successes was laid two or three decades ago in far-sighted policies adopted following World War II.

Focus will be brought to the debate over the role of the federal government in science and engineering education in the form of legislation sponsored by Chairman Fuqua establishing a National Coordinating Council on Technical, Engineering, and Scientific Manpower and Education. The problem of our society's declining scientific literacy will be with us beyond this year and Congressional action will most likely not be completed in this session or even with this legislation. I am hopeful that next year we will address the broader problems. The Congress, in authorizing funds for the National Science Foundation continues to give science education a far greater priority than does the Executive branch.

In the energy area, the debate over the federal role has been on the question of what point in the spectrum extending from basic research to the commercial application of research results, should government end its support. The Administration's view is that the government's role should end with basic research on the theory that involvement with applications would interfere with the "free market."

There are interesting exceptions to the Administration's view of the federal role which convey a more pragmatic philosophy. Clearly the Administration favors continued support for nuclear energy development over synthetic fuels, solar and alternative technologies. Demonstration projects in the nuclear area which clearly would not make it in the "free market" are supported by the Administration. The Clinch River Breeder Reactor in Tennessee is a good example of this. It is estimated that breeder technology will not be economically feasible until the middle of the next century at the earliest.

Energy is just one area in which I believe there is a desirable role for the government in providing a broad base of research along with development and demonstration in promising areas. The government can also play an important role in stimulating innovation and productivity. The U.S. still has the strongest research base in the world, and while this base is threatened, in my opinion, by the policies of the Reagan Administration, so too is our economy threatened today by our nation's competitive disadvantage resulting from insufficient support of new technologies.

Another area where the government can play an important role is in the area of Information Technology, an area on which your conference is focusing. Rapid advances in microelectronics and telecommunications technology, and the convergence of computers and telecommunications, have created new opportunities for economic growth, increased export markets, and gains in productivity. They will also permit increased public access to a variety of useful information as well as benefitting educational opportunities.

In the last Congress I introduced legislation calling for the establishment of a National Institute for Information Policy and Research. If we are to take advantage of the opportunities made possible by the new information technologies and be prepared for social changes while minimizing potential negative impacts, our social and governmental institutions must come to grips with the important information technology field.

I feel that the need for defining and analyzing national information policy issues is even more pressing now than it was in the last Congress, and for this reason I reintroduced the Information Science and Technology Act of 1981 last year. Issues of pressing concern include trans-border data flow of confidential information to American firms doing business abroad, direct broadcasting satellites, and increased competition from the development of targeted information strategy and policy in countries such as Japan, Great Britain, West Germany, and France. In the U.S., no comprehensive national effort has been undertaken to address the scientific, economic, and social issues arising from the rapid development of information technology and telecommunications, or to articulate national policies in light of this development. Information services provided by the private sector constitute an important and rapidly expanding part of our economy. However, no effective means currently exists to bring together public and private interests to discuss national information concerns in a cooperative forum.

I think such high-powered issues can not be left solely to the private sector if for no other reason than the international relations, social, and economic impacts of the new developments. Work in advanced science and technology, particularly in the area of information technology, is critical to meeting societal needs in such areas as environmental monitoring, epidemiology, agriculture, nutrition, and science education.

My own view is that the United States is suffering from schizophrenia over what the proper role of government is in supporting research, development, and demonstration. What programs do survive the "leave it to the free market" scrutiny end up overtaken by military priorities. It is not a good time for federal funding of science and technology. In some cases, as in education, we are cutting the very roots of our prosperity. I would like to conclude by quoting from Vannevar Bush's report to President Truman in 1945. He said:

"It has been a basic United States policy that government should foster the opening of new frontiers. It opened the seas to clipper ships and furnished land for pioneers. Although these frontiers have more or less disappeared, the frontier of science remains. It is in keeping with the American tradition--one which has made the United States great--that new frontiers shall be made accessible for development by all American citizens."

"Moreover, since health, well-being, and security are proper
concerns of Government, scientific progress is, and must be, of
vital interest to Government. Without scientific progress the
national health would deteriorate; without scientific progress we
could not hope for improvement in our standard of living or for an
increased number of jobs for citizens; and without scientific
progress we could not have maintained our liberties against
tyranny."

I hope we can heed those words today.

Thank you.

S.R. Heller and R. Potenzone, Jr. (Editors), *Computer Applications in Chemistry*
© 1983 Elsevier Science Publishers B.V., Amsterdam — Printed in The Netherlands

COMPUTER-ASSISTED STRUCTURE ELUCIDATION SYSTEM – FUNCTIONS THAT "CHEMICS" SYSTEM CAN PERFORM

SHIN-ICHI SASAKI and HIDETSUGU ABE
Toyohashi University of Technology, Tempaku, Toyohashi, JAPAN

The "CHEMICS" system, developed and published by the authors, is a computer-aided structure elucidation system for organic compounds. This article will describe the concepts on which CHEMICS has been developed and the future form into which it will be finally improved.

When a man is told of a computer-aided structure identification or determination method, he may be inclined to think of a method to retrieve a large file which has stored a majority of data. Such a method may be that if the infra-red spectral data of an unknown sample is entered into a system, the system will display relevant structures with the data similar to the input data after retrieval by data matching. In fact a number of data retrieval systems are now being utilized. However, in contrast with these systems, CHEMICS will analyze the input spectral data of an unknown compound, determine relevant substructures, combine them and display most probable structural formulas for the compound. Therefore, not a single structural formula is stored in a computer for CHEMICS system. We have often been disappointed at such questions from the audience who had observed our demonstration of CHEMICS system; "Well, how many structural formulas have been stored until now? Thirty thousand or fifty thousand?" The above explanation has been especially made to avoid such misunderstanding.

SYSTEMATIC APPROACH TO STRUCTURE ELUCIDATION

A surprising saver in time and labor for structure determination has been achieved by application of instrumental analysis beginning in the 1950's. A chemist can imagine a possible chemical structure for a compound, in consideration of functional

groups obtained from IR data, environments around H and C atoms in
the molecule from NMR data and skeletal structures from MS data.
Then he can draw a final conclusion by comparing the working
structure with the results of chemical experiments (eg. conversion
to derivatives, degradation into smaller molecules). Chemical
experiments can now be omitted for samples with some kinds of
molecular structures. However, in constructing a structure on the
basis of the information of instrumental analysis, little
difficulty will arise when the substructures to construct the
molecule do not overlap with each other. But this is not always
the case. Let us show a very simple example. If a sample is found
to have CH_3CO- and C_6H_5CO substructures, the structure may be
either $C_6H_5COCH_3$ or $C_6H_5COCOCH_3$. One may not decide which is the
correct structure unless other decisive information is available.
This example clearly shows that, for structure construction, we
must collect substructures which do not overlap with each other.
Consciously or unconsciously, one should have arrived at a correct
answer by taking the steps to avoid the overlapping. However, is
it always possible to arrive at a single solution if we obtain a
set of substructures that do not overlap with each other? Fig. 1
shows substructures of an N-acetyl derivative of actinobolamine
which Munk obtained from chemical and instrumental analysis of the
derivative. (ref. 1) After an elaborate effort to properly combine
the substructures, (A) - (D), one may have arrived at the
structure (1) as shown in Fig. 2. This may be the correct result
but a possibility still remains that there might be alternate
solutions. It would be quite natural to think of relying on a
computer with an extremely efficient capability for exhaustive
listing and enumeration as a tool to investigate the above
possibility. Therefore, Munk devised an algorithm to build
possible structures from the substructures listed in Fig. 1 and
showed that there are five possible structures (2 - 6) in addition
to the structure 1 as shown in Fig. 2, even under the limitations
derived from experimental results that (A) and (C) are linked
together and no double bond exists between carbon atoms. (Our
examination shows that there are 27 possible structures if the
substructures are allowed to be randomly linked.) If a man has hit
upon a most SIMILAR structure, he is least inclined to try to find
other possible structures. The computer compensates for this human
laziness. In other words, a computer-aided structure elucidation

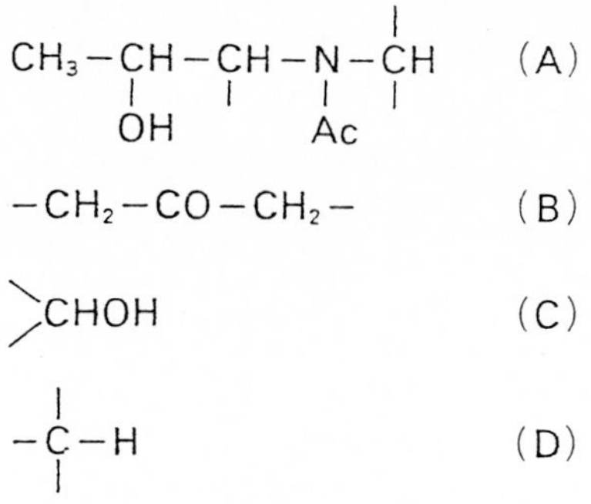

$$CH_3-CH-CH-N-\overset{|}{C}H \qquad (A)$$
$$\underset{OH}{|}\underset{Ac}{|}$$

$$-CH_2-CO-CH_2- \qquad (B)$$

$$\rangle CHOH \qquad (C)$$

$$-\overset{|}{\underset{|}{C}}-H \qquad (D)$$

Fig. 1. Substructures of N-acetyl-actinobolamine.

1 2 3 4 5 6

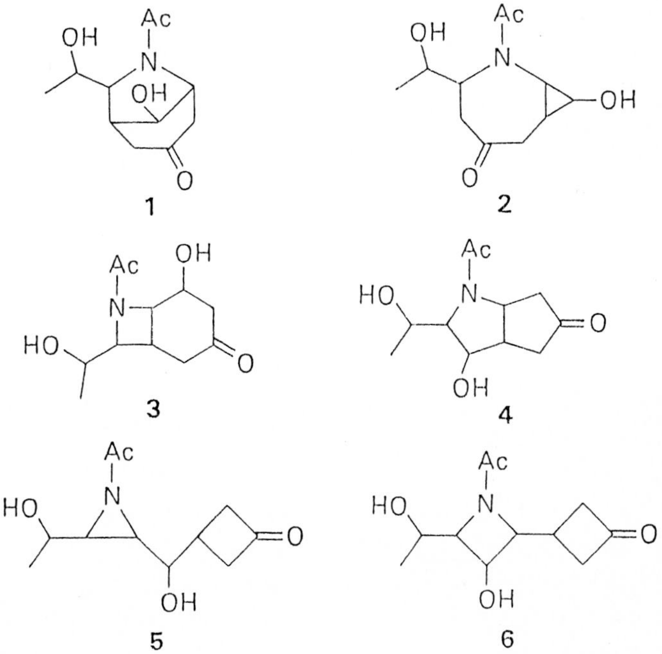

Fig. 2. Candidate structures afforded for N-acetyl-actinobolamine.

system is one in which a computer lists all the possible
structures that conform to all pieces of structure-relevant
information and selects the most probable structure from among
them. The paper of Munk was very suggestive in the sense that it
clearly points out such ideas for the structure elucidation
system. His concept led to the later development of "CASE", (ref.
2) structure elucidation system. We started the research of
CHEMICS (ref. 3) at about the same time as Munk. No significant
difference is found between the basic concept of our system and
that of Munk, which will be described later in this paper and was
already explained in Farumashia.(ref. 4) In addition, at the same
period, the first report was published on the structure
elucidation method which a Stanford group developed by application
of DENDRAL.(ref. 5) This is also common to our and Munk's systems
in the concept, "...lists all the possibilities...". From then
onward, system development was advanced of CASE, CHEMICS (ref. 6)
and DENDRAL (and later, CONGEN, (ref. 7) GENOA (ref. 8), in the
United States and Japan.

PREPARATION OF COMPONENTS IN CHEMICS SYSTEM

In short, structure elucidation may be that one thinks out all
possible structures that correspond to the obtained information,
lists them in front of one's eyes and then successively narrows
down candidate structures with reference to newly entered
information. Let us now consider the situation of a man who has
received a sample compound with no structure-relevant information
at all. (He has only received a sample without being informed of
any information about the structure, even of its past record.

If only the molecular formula of a compound is known, all the
possible structures for the compound may be listed from
consideration of the valencies of the constituent atoms, such as
4, 1, 2, 3,......for C, H, O, N.... Therefore, if a thorough
examination is made of all molecular formulas of homologs
belonging to various compound types, all the corresponding
structures may be listed, including the structures known to be
present, the structures which are still unknown but might be known
in the future and the structures which are least likely to exist
but take apparent forms of structures. For example, 753 structural
isomers, as shown below, may be calculated for hydrocarbons with
six carbon atoms.

C_6H_{14} 5 C_6H_6 217

C_6H_{12} 25 C_6H_4 185

C_6H_{10} 77 C_6H_2 85

C_6H_8 159   Total   753

Therefore, such a survey is made of compounds with carbon atoms up to 30, 0 atoms up to 5-6, and N atoms up to 3-4, and with compositions such as CH, CHN, CHNO....and all the resultant structural formulas are stored in files, it can be said that the file has stored all the structures of any samples that are to be objected to structure elucidation in the usual sense. The above-mentioned man with no information may be compared to a man who is at a loss what to do, standing against this gigantic file of structural formulas. However, it may be easily imagined that once he acquires necessary information, he will face to a file with surprisingly reduced number of formulas. Thus, when a structure elucidation system has received enough information in either quality or quantity, the system will complete the elucidation work and output a single correct solution that has survived.

CHEMICS may be said to have been designed to pick up the most suitable structure from this gigantic file, with the help of the information obtained from spectral data. Needless to say, however, it is not allowable to store such extraordinarily numerous structures (only for $C_{23}H_{48}$, 5,731,580!). Then, in place of these, CHEMICS is designed to store necessary and sufficient substructures (called components in the system) for building any structures. This set of components have been collected so as to construct any structure by selecting appropriate components from the set. CHEMICS has prepared 189 components for structure elucidation of organic compounds consisting of C, H and O atoms (Table 1). A trial and error method was adopted in the selection of these components, with due regard to the prerequisites that the components should have no overlapping part with each other and that the presence of the components can be deduced from spectral data. Moreover, 572 components have been newly prepared to handle the samples that contain N, halogen and S atoms in addition to C, H and O atoms (ref. 9) (Table 2).

Table 1   Components for compounds containing C, H and O

Code	Structure	Code	Structure	Code	Structure	Code	Structure
1~6	CH₃−C−(CH₃)(CH₃)	41~46	CH₃CH₂−	114	H−⟨'	142,143	>C=
		47~61	CH₃−CH<		H	144,145	=C=
7~12	−C−(CH₃)(CH₃)	62~88	−CH−	115	O=▷	146	⟨O'
		89~109	−CH₂−	116	−C≡CH	147~153	−O−
13~18	CH₃−C−	110	CH₂=	117,118	−CH=	154~174	−CO−
19~25	(CH₃)(CH₃)CH−	111	CH₂ (ring)	119~124	−OCHO	175	=C=O
				125~127	−OH	176	−C≡C−
26~31	CH₃O−	112	H (ring)	128~133	−COOH	177~182	−⟨'
32~34	CH₃−	113	HO−⟨'	134~140	−CHO		
35~40	CH₃CO−			141	O=◁	183~188	−C−

† Components for forming aromatic ring

Table 2   Components for compounds containing
            C,H,O,N,S and halogens

Code	Structure	Code	Structure
1~8	CH₃−C−(CH₃)(CH₃)	357~364	−N(=O)(→O)
64~71	CH₃S−	473~484	S(=O)(→O)
189~196	−C≡N	569	−F
		570	−Cl
259~266	−CH=N⁺=N⁻	571	−Br
		572	−I

CHEMICS METHOD FOR STRUCTURE ELUCIDATION

CHEMICS has the following tasks:

1 to select necessary components from the prepared component set, with reference to chemical spectra.

2 to generate structural formulas from the selected components.

3 to exclude inappropriate structures from the generated structures.

4 to output candidate structures as few as possible, ideally one and correct solution.

The authors will explain how these tasks are solved in CHEMICS, by referring to the CONCEPTUAL diagram of CHEMICS, as shown in Fig. 3.

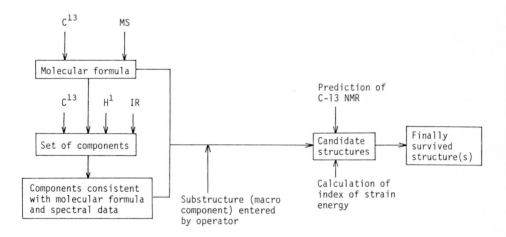

Fig. 3. Conceptual diagram of CHEMICS method.

Molecular Formula as an Indispensable Information

The number of components can be remarkably reduced if the molecular formula of a sample is known. For example, all the components related with O atom can be discarded if a molecular formula is found to contain no O atom. Thus a molecular formula is an indispensable input data in CHEMICS and moreover, it also plays an important role in structure generation as will be described later. A man may input a molecular formula which he has acquired from an independent source. Also, CHEMICS can AUTOMATICALLY predict molecular formula from MS and C-13 NMR data. A computer investigates molecular ion clusters (several peaks around a

molecular ion peak) in MS spectrum and determines a molecular
weight. Then, it computes all the possible compositions from the
molecular weight. Every composition is successively compared with
the C-13 NMR data and, then the consistent compositional formulas
with the data are adopted as the molecular formulas. When it is
impossible to narrow down to a single formula, formulas are ranked
by comparison of simulation results with observed values. For
example, a certain sample is found to have an ion cluster
consisting of four peaks in the highest mass region of the MS

```
M/Z    REL.INT. 0_____10_____20_____//___100  (%)
198.0    1.5     **
199.0     .7     *
200.0    1.7     **
201.0     .6     *
202.0     .2

213.0  100.0     ****************************** ****
214.0   10.7     **********
215.0   98.3     ****************************** **
216.0    9.9     **********

228.0   25.5     ****************************
229.0    2.9     ***
230.0   25.9     ****************************
231.0    2.8     ***
```

```
CLUSTER  NO.1     M/Z      R.A
                  228.0    25.5
                  229.0     2.9
                  230.0    25.9
                  231.0     2.8
```

Fig. 4. Mass spectrum and molecular ion cluster of a certain
        compound.

spectrum (Fig. 4). When a computer is informed that the sample
does not contain N atoms, it points out ions with m/z values of
228 and 230 as the candidate molecular ions and calculates
possible composition for each of them (Fig. 5). If the sample is
assumed to have such C-13 NMR spectrum as shown in Fig. 6, the
number of candidate formulas is limited to 3 for m/z 230 and 2 for
m/z 228, since the allowable minimum values are 8 and 7 for C and
H, respectively (Fig. 7). Then simulation-baedd calculation of
discrepancies gives high priority to $C_{10}H_{13}OBr$ and $C_9H_9O_2Br$ Fig.8.

```
CANDIDATE
     M/Z    R.A
    230.0   25.9
    228.0   25.5
```

CALCULATION OF MOLECULAR FORMULA

ELEMENTS		C	H	O	N	F	CL	BR	I	SI	P	S
M/Z=230	NO.											
	1	3	3	7	0	0	0	1	0	0	0	0
	2	4	7	6	0	0	0	1	0	0	0	0
	3	5	11	5	0	0	0	1	0	0	0	0
	4	7	3	4	0	0	0	1	0	0	0	0
	5	8	7	3	0	0	0	1	0	0	0	0
	6	9	11	2	0	0	0	1	0	0	0	0
	7	10	15	1	0	0	0	1	0	0	0	0
	8	11	3	1	0	0	0	1	0	0	0	0
	9	3	4	2	0	0	0	2	0	0	0	0
	10	4	8	1	0	0	0	2	0	0	0	0
M/Z=228	NO.											
	1	3	1	7	0	0	0	1	0	0	0	0
	2	4	5	6	0	0	0	1	0	0	0	0
	3	5	9	5	0	0	0	1	0	0	0	0
	4	6	13	4	0	0	0	1	0	0	0	0
	5	7	1	4	0	0	0	1	0	0	0	0
	6	8	5	3	0	0	0	1	0	0	0	0
	7	9	9	2	0	0	0	1	0	0	0	0
	8	10	13	1	0	0	0	1	0	0	0	0
	9	11	1	1	0	0	0	1	0	0	0	0
	10	3	2	2	0	0	0	2	0	0	0	0
	11	4	6	1	0	0	0	2	0	0	0	0

Fig. 5. Possible elemental compositions of candidate molecular ions, m/z 230 and 228.

```
C13-NMR SPECTRUM

    SIGNAL NUMBERS  8
    MULTIPLICITY  S 3
                  D 3
                  T 0
                  Q 1
```

Fig. 6. Signal number and multiplicity of C-13 NMR of the sample whose mass spectrum is shown in Fig. 4.

		C	H	O	N	F	CL	BR	I	SI	P	S
M/Z=230	NO.											
	1	8	7	3	0	0	0	1	0	0	0	0
	2	9	11	2	0	0	0	1	0	0	0	0
	3	10	15	1	0	0	0	1	0	0	0	0
M/Z=228	NO.											
	1	9	9	2	0	0	0	1	0	0	0	0
	2	10	13	1	0	0	0	1	0	0	0	0

Fig. 7. Candidate molecular formulas supported by C-13 NMR.

SIMULATION MASS SPECTRUM

```
[1]    C10 H13 0 BR  (MW.228)   D=   .4
[1]    C9  H9  02BR  (MW.228)   D=   .4
[3]    C8  H7  03BR  (MW.230)   D= 12.8
[3]    C9  H11 02BR  (MW.230)   D= 12.8
[3]    C10 H15 0 BR  (MW.230)   D= 12.8
```

Fig. 8. CHEMICS adopts the molecular formulas with lower
discrepancies after simulation: $C_{10}H_{13}OBr$ and $C_9H_9O_2Br$.

The series of procedures described above are automatically carried
out in CHEMICS, only when MS and C-13 data are entered. This may
be found that CHEMICS performs quite properly to predict the
molecular formula as shown in Table 3 listing the test results of
molecular formula determination for various compounds.

Screening of Components by Spectral Data
    Among the components which have survived because of their
consistency with the molecular formula determined as above, some
will be further discarded which are inconsistent with subsequently
entered H-1, C-13 NMr or sometimes IR data. The range of
occurrence was determined for H-1 shifts of H-containing
components and for C-13 shifts of C-containing components, by

Table 3   Predicted molecular formulas for various compounds

Sample compound	Molecular formula	Number of answers
3-Pyridine carboxyamine	$C_6H_6ON_2$	1
Benzenesulfonic acid	$C_6H_6O_3S$	4
1,3-Dimethyluracil	$C_6H_8O_2N_2$	1
n-Butylthioacetate	$C_6H_{12}OS$	1
Benzenesulfonyl chloride	$C_6H_5O_2Cl_2S$	3
4-Hydroxy-3-methoxybenz-aldehyde	$C_8H_8O_3$	1
n-Octylmercaptane	$C_8H_{18}S$	2
1,3,3-Trimethylbicyclo [2,2,1]heptan-2-one	$C_{10}H_{16}O$	1
n-Butylbenzoate	$C_{11}H_{14}O_2$	1
Azobenzene	$C_{12}H_{10}N_2$	5
m-(1-hydroxy-1-methyl-ethyl)cumene	$C_{12}H_{18}O$	1

The numeral 1 means that a single correct formula has been obtained. The numeral 4 means that three candidate formulas survive in addition to the correct answer. This applies corresponding to another numerals.

Table 4   Component-chemical shift (ppm) correlation table

Code	Structure	^{1}H NMR	^{13}C NMR
1	CH_3	84.0 ~ 60.0	26.02~31.13
2	CH_3-C-	96.0 ~ 72.0	24.47~33.57
3	CH_3	90.0 ~ 60.0	25.48~34.04
⋮		⋮ ⋮	⋮ ⋮
10	CH_3	90.0 ~ 48.0	14.72~36.27
11	$-C-$	90.0 ~ 48.0	10.12~36.27
12	CH_3	89.4 ~ 36.0	6.80~32.61
⋮		⋮ ⋮	⋮ ⋮
106		220.0~108.0	20.46~56.42
107	$-CH_2-$	162.0~102.0	18.88~53.09
108		180.0~114.0	13.27~54.14
109		144.0~ 0.0	14.66~57.16
⋮		⋮ ⋮	⋮ ⋮
153	$-O-$	— —	⋮ —
⋮			

reference to the data so far obtained. Those ranges are tabulated
into correlation table after some statistical treatment (Table 4).
The selection of components by spectral data in CHEMICS simply
means that the spectral data of a sample is compared with the
component-chemical shift relation table and the components
consistent with the data are left to be reserved. In practice, the
selection is not a single matching but a more sophisticated
process because various parameters influence the selection in a
complicated way. The detail was described in another paper.(ref.
10)

Structure Building from Selected Components

     The next step is to build up structures by use of the
components which have been selected as being not contradictory to
a molecular formula and spectral data. Since such principle comes
into play in component selection as to discard only the components
that are contradictory to data, even such components as are picked
up may seem to be useless from a chemist's eyes. However, this
situation should not be thought as inevitable but it is rather
natural in view of the mission of this system never to miss the
correct solution. Table 5 shows the components which have been
selected with reference to the spectral data of a certain sample.
In the first step to build structures, the molecular formula plays
an important role. The structural formula to be built should have
the same kinds and numbers of atoms of the molecular formula.
Therefore, whether how many of which components are equal to the
molecular formula is examined to 29 components in Table 5. It is a
favorite function of a computer to list all the possible
combinations. CHEMICS does not collect all the random combinations
but involves a logic to exclude prohibited combinations when
coordination-prohibited component pairs are found in view of NMR
data. Fig. 9 shows five combinations of components chosen from 29
components in Table 5, from the assumption that the molecular
formula is $C_9H_{14}O$. There are no other combinations than the five.
The next step is to generate structures from individual component
combinations. The generation is carried out without leaving a
single possibility undone, in due consideration of the principle
that most of the components have limited species to link (as noted
in the footnote of Table 5). On the basis of a specially designed
logic, Connectivity Stack, (ref. 11) the system functions properly

	Code	Structure			Max.
1	10	*gem*-(CH₃)₂—	(D)		1
2	11	*gem*-(CH₃)₂—	(T)		1
3	12	*gem*-(CH₃)₂—	(C)		1
4	14	CH₃— Ċ —	(Y)		2
5	17	CH₃— Ċ —	(T)		2
6	33	CH₃—	(D)		2
7	38	CH₃CO—	(D)		1
8	40	CH₃CO—	(C)		1
9	106	—CH₂—	(C)	(K)	2
10	107	—CH₂—	(C)	(D)	2
11	108	—CH₂—	(C)	(T)	2
12	109	—CH₂—	(C)	(C)	2
13	118	—CH=			1
14	143	>C=			1
15	144	=C=			1
16	145	=C=			1
17	146	—O—			1
18	153	—O—			1
19	172	—CO—	(C)	(D)	1
20	173	—CO—	(C)	(T)	1
21	174	—CO—	(C)	(C)	1
22	175	O=C=			1
23	177	Y	(O)		1
24	182	Y	(C)		1
25	184	C	(Y)		1
26	185	C	(K)		1
27	186	C	(D)		1
28	187	C	(T)		1
29	188	C	(C)		1

TABLE   5. Components selected from Table   1   with reference to the spectral data of a certain sample and the maximum number of components that they may exist.

not to reproduce the same structure and not to fail to generate the structures to be justly built up. This logic is important but not explained here since it has been described elsewhere. Fig. 10 shows 12 structural formulas obtained from five combinations of components. Although improbable formulas may be involved, we should treat them all as having exactly the same possibility at this time, so long as they are all incontradictory with the molecular formula and spectral data.

Means to Reduce Answers

Input of Human Chemical Knowledge(ref. 12)

A man often has some information about the structure of a sample. This may be obtained from the past record of the sample or the experience in its laboratory handling. One should make the most of such information. Now, if 29 components are obtained for a sample with the molecular formula, $C_9H_{14}O$ (Table 5), the system

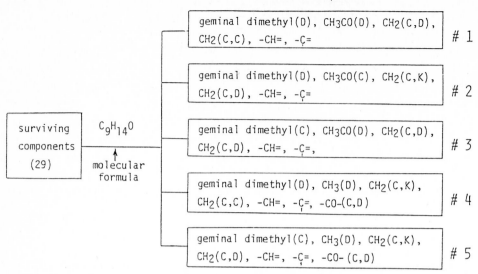

Fig. 9. Combinations of components selected from Table 5, so as to
be equal to $C_9H_{14}O$.

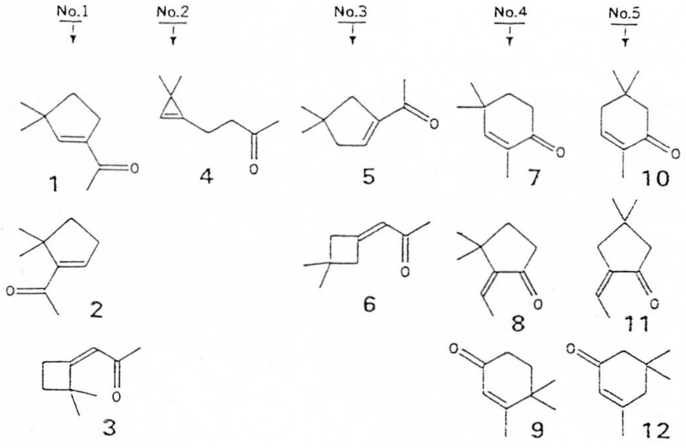

Fig. 10. Structures generated from No.1-5 component combinations
in Fig. 9.

will automatedly output twelve possible structures (Fig. 10). At
this time, if a man has been convinced that the sample should have

a substructure of $CH_3-\overset{|}{C}=CH-CO-$, he may do well to enter the information into the system. The entered substructure is "degraded" to components as shown in Table 1, which are then compared with the components that the system has selected. The system will adopt the entered information, only when the former components are involved in the later components. This means that the components which the system has selected with a full safety factor should take preference over the information which a man has entered. Since the degradation products of $CH_3-\overset{|}{C}=CH-CO-$ coincide with the components with code numbers of 33, 143, 118 and 172 in Table 5, the substructure is adopted. Furthermore, the substructure, $Ch_3-\overset{|}{C}=CH-CO-$, can not be generated from No. 1, 2 and 3 combinations which are therefore, not used for structure construction. Only two structures, 9 and 12 are generated from No. 4 and 5 combinations and will be displayed as the whole answer. Although a man may input any substructures to reduce output structures, the system is designed not to give credence to all the information that a man has entered.

Reduction of Structures by Prediction of C-13 NMR Data (ref. 13)

If prediction can be made of the spectral data for a multiple of candidate structures which are displayed, the structures may be reduced by comparison between predicted and observed data. It is not always easy to predict the entire spectrum from a structural formula. Therefore, prediction is made of only the number of C-13 signals, which is to be compared with that of a sample.

The number of C-13 signals of a compound should be equal to the number of non-equivalent carbon atoms in the molecule. To discriminate between equivalent and non-equivalent C atoms, the connectivity stack which can make unique expression of structures is utilized. When no change is observed in the stack expression even if any pairs of C atoms are exchanged with each other, the two carbon atoms can be recognized as equivalent. When numerals are allocated to individual carbon atoms of methylcyclobutane as shown in Fig. 11, the connectivity stack is expressed by S_O : 1100110001, because the stack expression is made by the numeral series made from the upper-half triangle matrix in the atom-atom connectivity table (1 denotes linkage, and 0 non-linkage in Fig. 11). Exchange of C-2 for C-3 (2 of Fig. 11) and of C-1 for C-2 (3 of Fig. 11) will give the stack expression of S_1 and S_2,

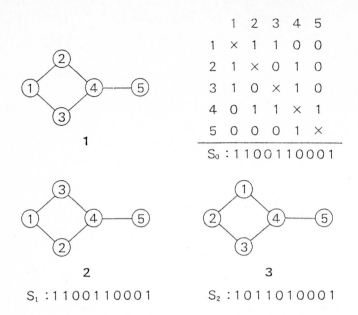

```
              1  2  3  4  5
           1  ×  1  1  0  0
           2  1  ×  0  1  0
           3  1  0  ×  1  0
           4  0  1  1  ×  1
           5  0  0  0  1  ×
           ─────────────────────
           S₀ : 1 1 0 0 1 1 0 0 0 1
```

$S_0 : 1100110001$

$S_1 : 1100110001$ $S_2 : 1011010001$

Fig. 11. Structural expressions for methylcyclobutane.

respectively. Because $S_0 \neq S_1$, C-2 and C-3 are equivalent. Because $S_0 = S_2$, C-1 and C-2 are non-equivalent. The stack expressions for exchanges between all the possible pairs of C atoms will be found to be different from S_0 except for exchange between C-2 and C-3 atoms. Therefore, the chemical shifts of C-1, (C-2, C-3), C-4 and C-5 should be different and the number of signals is predicted to be four. Among those carbon atoms that are judged to be topologically equivalent, some may indicate different chemical shifts because they are not equivalent from the chemical point of view. For example, although 3,3,5-trimethylcyclohexanone is predicted to have eight signals by the above-mentioned method, nine signals are experimentally observed. The reason is that two methyl groups at C-3 are in different chemical environment. Therefore, some allowance should be provided for the prediction method, but the method is sufficiently usable. For example, let us suppose indan to be an unknown sample. When the H-1 and C-13 NMR data of the sample are entered, CHEMICS will output four candidate structures (Fig. 12). And the predicted number of C-13 signals are 5, 6, 4 and 4 for structures a, b, c and d, respectively. Since the signal number is found to be five from the input spectral data, structure (a) is adopted as the sole correct answer. Table 6

Table 6   Reduction of candidate structures by
          C-13 NMR prediction

Sample compound	Number of structures generated by spectral analysis	Results after C-13 NMR prediction
Indan	4	1
Benzil	20	1
Azulene	18	5
Camphor	32	26
2-Cyclohexyl-cyclohexanone	147	63
2,3-Dimethyl-naphthalene	273	19

shows the effectiveness of this predictive method for those
compounds to which CHEMICs have given relatively many candidate
structures by use of the spectral data alone.

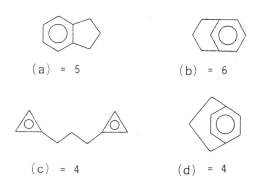

(a) = 5   (b) = 6

(c) = 4   (d) = 4

Fig. 12. Predictive number of C-13 NMR signals for four
         structures.

Reduction of Candidate Structures by Index of Strain Energy

One of the reasons why CHEMICS provides numerous candidate
structures for a sample is that it outputs even those structures
that are unreasonably highly strained unless they are inconsistent
with the spectral data. Such an attitude should be avoided as to

exclude such structures thoughtlessly, merely because they are
impossible from the view point of chemistry. However, it may be
necessary to give a sort of comment to extremely strained
structures, suggesting that they might be impossible.

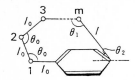

Fig. 13. Calculation of strain.

Structure straining is particularly high in unsaturated
(small-membered) ring structures, let along cage structures.
Therefore, we have made a sort of strain level for such
structures, in order to make a rough guess as to their existence
possibility. The method to calculate the strain level will be
illustrated below. The compound, (m)paracyclophane, consists of a
benzene ring and an unsaturated (m+4)-membered ring. In this sort
of structure having bicyclic condensed rings, the more unsaturated
ring (benzene ring, here) is set as the basis and the carbon chain
starting at one juncture and ending at another is taken into
consideration (1 to m in Fig. 13). Give the standard bond angle
(θ_0) and bond length (Z_0) to sp, sp^2 and sp^3 bonds and stretch a
chain so that individual carbon atoms may take ordinates as close
as possible towards another juncture. By use of the bond angles,
θ_1 and θ_2, and the bond length, Z, obtained lastly calculate the
strain index, E_T, in accordance with the following equations.

$$E_s = K_s \ [(Z - Z_0) \diagup n]^2 \ \text{x} \ n$$

$$E_\theta = K_\theta \ [(\frac{\theta_1 - \theta_0}{n}) + (\frac{\theta_2 - \theta_0}{n})^2] \ \text{x} \ n$$

$$E_T = \underset{i \ \text{to} \ M}{\text{sum}} \ (E_{si} + E_{\theta i})$$

Strain is assumed to be non-existent when $l < l_0$.

E_s: strain for bond length

E_θ: strain for bond angle

n : number of ring members   M: number of rings in structure

K_s, K_θ: constants determined to give the highest correlation between observed and calculated E_T values, by use of E_T of a molecule with the strain energies experimentally determined

Fig. 14 shows structural formulas which CHEMICS has generated for a certain sample. Calculated E_T values for structures 1, 2, 3 and 4 are 12, 141, 13 and 12, respectively. This suggests that structure 2 might not be possible in terms of strain energy.

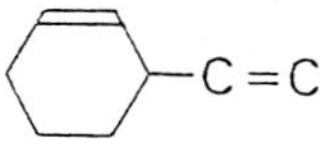

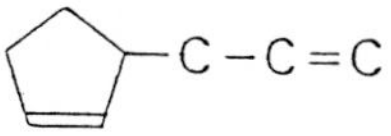

Fig. 14. Index of strain for some unsaturated ring structures.

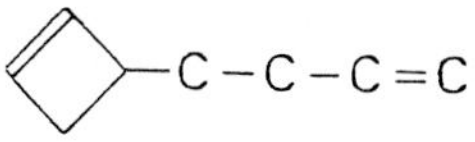

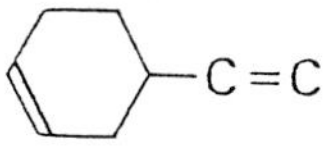

RESULTS

Table 7 lists final results of 21 sample compounds, which CHEMICS has obtained after analysis of the input spectral data and a man's relevant information and screening by C-13 NMR prediction and strain energy calculation. One can recognize that the correct solution has always been obtained for any sample. In addition, the table shows what kind of candidate structures have survived.

Table 7   Final outputs of CHEMICS for various compounds

Left: correct structure
Right: Candidates simultaneously generated

Results of N-containing samples

Ethylpyridine	C_7H_9N	3	ACH x 4
Diisobutylamine	$C_8H_{19}N$	$\begin{cases}15\\1\end{cases}$	-NH(CS) x 1
2,6-Diethylaniline	$C_{10}H_{15}N$	6	$NH_2(Y)$ x 1
Benzyl isopropylamine	$C_{10}H_{15}N$	1	ACH x 5
Ethylmorpholine	$C_6H_{13}NO$	10	-N< x 1

(Appendix of Table 7)

REFERENCES

1   M.E. Munk, C.S. Sadano, R.L. McLean, and T.H. Haskel, J. Am.
    Chem. Soc., 89 (1967) 4158.

2   C.A. Shelley, H.B. Woodruff, C.R. Snelling, and M.E.Munk, in
    D.H. Smith (Ed), "Computer Assisted Structure Elucidation", ACS
    Symposium Series 54, Am. Chem. Soc., Washington, D.C. (1977)
    92.

3   S. Sasaki, H. Abe, T. Ouki, M. Sakamoto, and S. Ochiai, Anal.
    Chem., 40 (1968) 2220.

4   S. Sasaki, Farumashia, (in Japanese) 7 (1972) 440.

5   A.M. Duffield, A.V. Robertson, C. Djerassi, B.G. Buchanan, C.L.
    Sutherland, E.A. Feigenbaum, and J. Lederberg, J. Am. Chem.
    Soc., 91 (1969) 2679.

6   S. Sasaki, H. Abe, Y. Hirota, Y. Ishida, Y. Kudo, S. Ochiai, K.
    Saito, and T. Yamasaki, J. Chem. Info. Comp. Sci., 18 (1978)
    211.

7   R.E. Carhart, D.H. Smith, H. Brown, and C. Djerassi, J. Am.
    Chem. Soc., 97 (1975) 5755.

8   R.E. Carhart, D.H. Smith, N.A.B. Gray, J.G. Nourse, C.
    Djerassi, J. Org. Chem., 46 (1981) 1708 D.H. Smith, N.A.B.
    Gray, J.G. Nourse, and C.W. Crandell, Anal. Chim. Acta, 133
    (1981) 471.

9   S. Sasaki, H. Abe, I. Fujiwara, T. Okuyama, T. Nishimura, The
    6th International Conference on Computers in Chemical Research
    and Education, Washington, D.C. (1982).

10 S. Sasaki, H. Abe, Y. Hirota, Y. Ishida, Y. Kudo, S. Ochiai,
    K. Saito and T. Yamasaki, Sci. Rep. Tohoku Univ., 60 (1978)
    154. T. Yamasaki, H. Abe, Y. Kudo and S. Sasaki, in D.H. Smith
    (Ed.), Computer Assisted Structure Elucidation, ACS Symposium
    Series 54, Am. Chem. Soc., Washington, D.C., (1977) 108.

11 Y. Kudo and S. Sasaki, J. Chem. Doc., 14 (1974) 200.

12 S. Sasaki, I. Fujiwara and T. Yamasaki, Anal. Chim. Acta 122
    (1980) 87.

13 I. Fujiwara, T. Okuyama, T. Yamasaki, H. Abe and S. Sasaki,
    Anal. Chim. Acta, 133 (1981) 527.

S.R. Heller and R. Potenzone, Jr. (Editors), *Computer Applications in Chemistry*
© 1983 Elsevier Science Publishers B.V., Amsterdam — Printed in The Netherlands

COMPUTERS IN CHEMICAL RESEARCH AT THE NOVOSIBIRSK SCIENTIFIC CENTRE

V. A. KOPTYUG

Novosibirsk Scientific Centre of Siberian Branch of the USSR Academy of Sciences

The Novosibirsk Scientific Centre currently incorporates 21 institutes conducting research in all of the main fields of natural sciences. This has created favourable conditions for comprehensive research work on problems lying at the junction of different sciences, including those which demand the work of mathematicians and specialists in computer science. Therefore, it is natural that the Novosibirsk Scientific Centre and, more recently, of other centres of Siberian Branch should successfully develop the so-called "mathematical technology" for the various fields of science and technique. That term implies mathematical description of any phenomenon, process or properties, etc. and their computer simulation, using the available data banks when appropriate. Such work has been extensively carried out in the field of chemistry. The further examples will give a general idea of the trends of computerization of chemistry and the related sciences at the Novosibirsk Scientific Centre.

Experiment control in the field of measurement of various properties of substances and their interconversions is of great importance for chemistry as well as for any other branches of science. This task is considerably facilitated by the works of the Institute of Automatics and Electrometry (the head is Academician Yu. Ye. Nesterikhin). The Institute is developing methodology and hardware for computerized systems designed for experimental research. In the early 70's, the Institute of Automatics and Electrometry recommended other institutes of the Novosibirsk Scientific Centre to use the module principle in designing computerized systems on the basis of the CAMAC

standard (Computer Application to Measurement, Acquisition and
Control), widely used in experimental physics. The Institute has
developed dozens of modules for various purposes (Fig. 1) and
has put them into production. This has sharply accelerated and
unified the work on computerization of scientific experiments at
the institutes of the Novosibirsk Scientific Centre, including
the chemical and biological ones. Thus, a number of computerized
systems have been designed: those for testing the properties of
catalyzers (The Institute of Catalysis), for recording and
processing molecular spectra (The Institute of Organic
Chemistry), for experimental research work in biology (The
Institute of Automatics and Electrometry), etc.

Fig. 1. CAMAC modules.

It is interesting (being natural at the same time) that
extended work on computerization of measurement has entailed
increased requirements to measurement instruments and involved
scientific workers in designing such instruments. Thus, at the
Molecular Biology Department of the Novosibirsk Institute of
Organic Chemistry (the head of the department is Academician D.
G. Knorre), a high performance liquid chromatograph "Ob-4" was
created with optical detection at several wavelengths in the
range of 190-360 nm (Fig. 2). Computer processing of measurement
results and some other features of its design allow to take for

analysis the amount of compound 10 times as small as that
required for the best instruments sold in the world market. The
multiwave detection and the on-line mode give it a number of
important advantages.

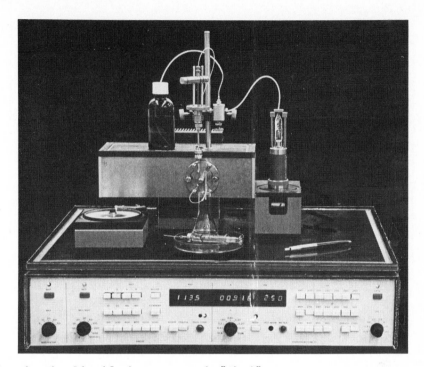

Fig. 2. The liquid chromatograph "Ob-4".

These are:
- optical density ratio control at different wavelengths of
  chromatographic peak allow to make conclusions as to its
  homogeneity;
- high precision in determination of that ratio facilitates
  identification of compounds related to each peak;
- the use of optical density ratios at several wavelengths
  facilitates resolution of unresolved peaks.

Creation of that instrument has raised the level of works on
ultramicroanalysis in molecular biology, clinical chemistry,
environmental pollution control, etc. (this work is being done
under the guidance of Dr. M. A. Grachov). One of the examples to
this is method for the determination of nucleotide sequences in
oligonucleotides by analysing their partial enzymic

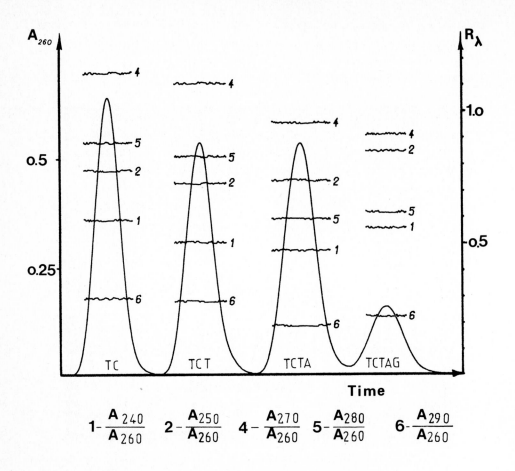

$$1 - \frac{A_{240}}{A_{260}} \quad 2 - \frac{A_{250}}{A_{260}} \quad 4 - \frac{A_{270}}{A_{260}} \quad 5 - \frac{A_{280}}{A_{260}} \quad 6 - \frac{A_{290}}{A_{260}}$$

Oligonucleotide		$Q \cdot 10^5$			
n	n–1	A	G	C	T
pTpCpTpApG	pTpCpTpA	1200	2	1200	800
pTpCpTpA	pTpCpT	50	2550	2550	4570
pTpCpT	pTpC	190	170	110	10
pTpC	pT	2000	900	400	2000

STRUCTURE: (T)CTAG

Fig. 3. The chromatogram of enzymatic decomposition products of an oligonucleotide identified by computer as pTpCpTpApG.

decomposition products, developed at the Institute of Organic Chemistry (Fig. 3).

Extensive work is carried out at the Novosibirsk Scientific Centre on computerization of chemical and the related research with the aid of large data banks. Thus, the Scientific-Information Centre on Molecular Spectroscopy (the head is Academician V. A. Koptyug) has molecular spectroscopy data banks which contain IR and UV spectra, mass-spectra, ^{1}H and ^{13}C NMR spectra of dozens of thousands of organic compounds in computer readable form.

The appropriate software and hardware enable one to work with spectroscopic data banks in the interactive mode both from near and distant terminals. During the last years, these data banks have been effectively used to solve the following problems:

1. Search for compounds absorbing or transmitting within certain ranges of spectra (light filters, laser media, photochemically active compounds, etc.).
2. Identification of known organic compounds by matching their spectra with reference spectra (components of industrial compositions, criminalistic problems, environmental pollution control, etc.).
3. Elucidation of the structures of unknown compounds by searching for compounds with similar spectral characteristics.

To create data banks containing spectra recorded on the on-line spectrometers certainly does not present any difficulties. Problems arise in digitizing previously published spectra issued as atlases and card files. For the purpose of input of graphic spectral data in a computer, a special semi-automatic device has been designed and is used at the Scientific Information Centre on Molecular Spectroscopy (the "Contour" device, Fig. 4).

Elucidation of the structures of unknown compounds with the aid of spectral information data banks has proved to be the most interesting problem. This procedure is based on recognition of large structural fragments of molecules. Practice has shown that compounds with similar spectral characteristics appear to be

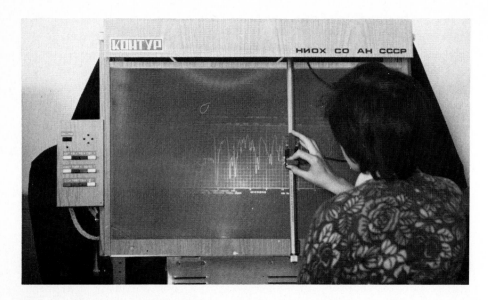

Fig. 4. The device for input of spectral curves.

close structural analogues of the compound in question (Fig. 5).

It is essential that a computer gives solutions to such problems even in the cases when the corresponding spectrum-structure relationships are not quite clear for researcher. Thus, it helps to obtain structural information from mass-spectra (Fig. 6), interpretation of which has always been difficult for a chemist.

It is evident that this approach opens up possibilities for the creation of computerized systems for solving the problems of direct (structure--→ spectra) and reverse (spectra--→ structure) relationships, as applied to chemistry. This work has been done by us during recent years. In solving these problems, we have encountered the problem common to all computerized systems of the "structure-property" type: the problem of input and subsequent handling of chemical structures.

The first part of this problem has found its solution in a semi-automatic device for encoding and input of chemical structures designed by G. P. Ulyanov and A. R. Maslov, (the

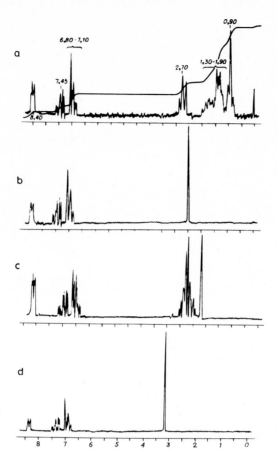

Fig. 5. PMR spectra (60Mc) of an investigated compound of $C_{10}H_{15}N$ composition (a) and of compounds selected by computer: 2-methylpyridine   (b), 2-B-dimethylaminopyridine   (c),   1,2-bis-(d-pyridyl)-ethane (d).

"Graph" device, Fig. 7). To input structural formulas, the operator is simply to draw them with a special electronic ball-point pen on a printed form lying on the tablet of the device. The software developed by V. N. Piottukh-Peletsky transforms the signals to a connectivity matrix containing the description of the structure. To solve the second part of the problem, that is to develop the software for handling structural information, we have made use of the works of the Institute of Mathematics on the graph theory (under the guidance of Prof. V.

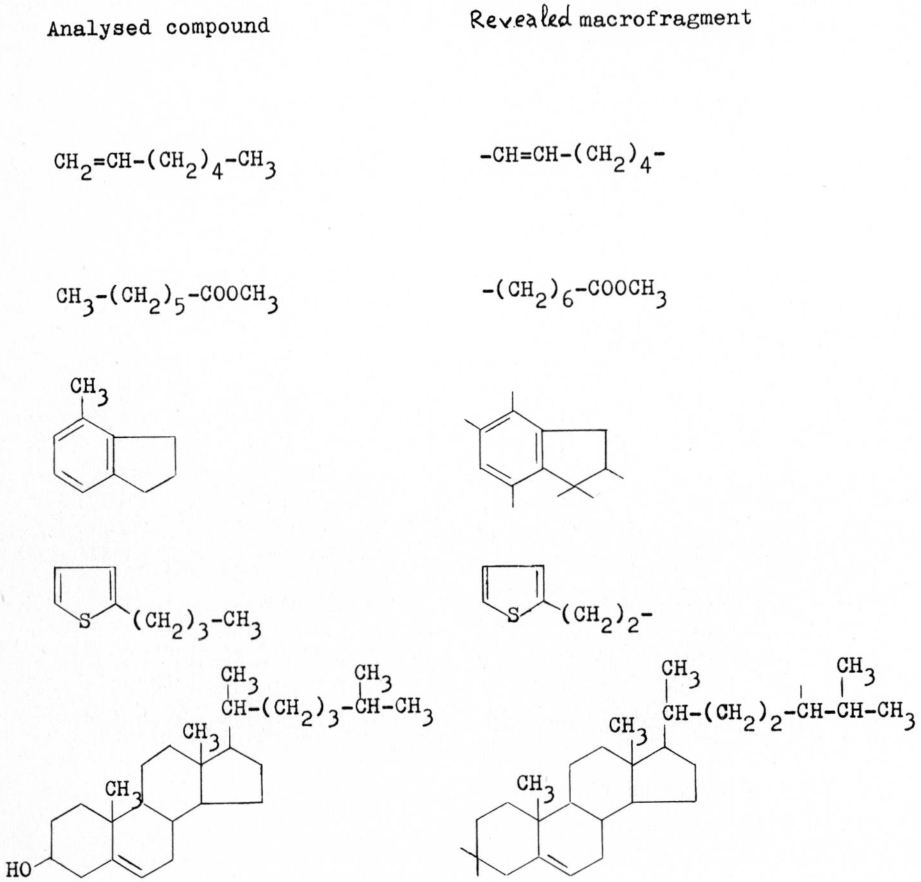
Analysed compound
Revealed macrofragment
CH2=CH-(CH2)4-CH3
-CH=CH-(CH2)4-
CH3-(CH2)5-COOCH3
-(CH2)6-COOCH3
CH3
S
(CH2)3-CH3
S
(CH2)2-
CH3
CH3
CH-(CH2)3-CH-CH3
CH3
CH3
HO
CH3
CH3
CH-(CH2)2-CH-CH-CH3
CH3
CH3

Fig. 6. Examples of large structural fragments recognized by comparison of structures selected by computer according to the mass-spectrum of compound under investigation.

Skorobogatov).

Today the methodology has been developed for designing "Graph" - based systems for encoding large volumes of structural information, as well as instrumentation for these systems (Fig. 8). The encoding output of each device is 20 to 25 structures per hour. This has created the possibility of quick formation of structural data bases for various data banks of the "structure-property" type, employed in a number of fields of scientific research work.

Fig. 7. The "Graph Device".

It should be noted that this system provides for the output of structural formulas in the form customary for a chemist (Fig. 9). This is of great importance while preparing various spectral data publications.

The creation of computerized physical chemical data banks and their employment meets certain difficulties connected with digitization of graphic information: spectral curves, phase diagrams, etc. In principle, plots may be encoded at a very small pitch and displayed with the required precision. But that results in a greater size of memory and increased machine time

Fig. 8. The "Graph" based system for encoding large volume of
structural information.

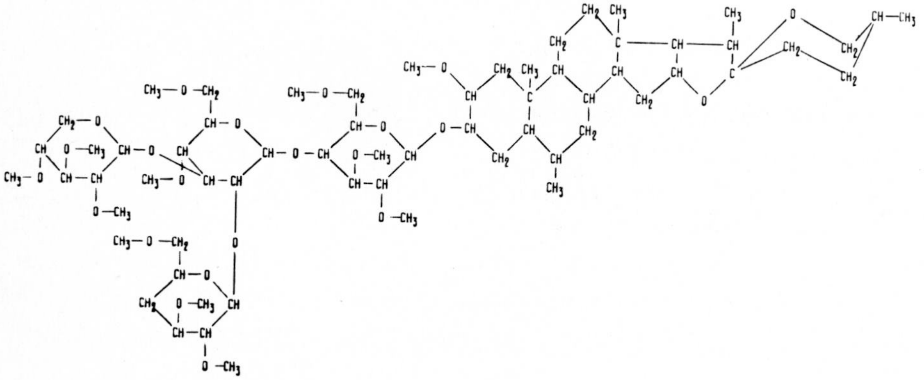

Fig. 9. An output of structural data.

required for data survey. One of the ways to overcome these
difficulties is to create two computer files, one of them
containing a complete set of graphical data and another - only
specific points. The second, abbreviated, file is intended for
the detailed analysis of graphical information and its
visualization.

Fig. 10. The system of holographic data storage and retrieval.

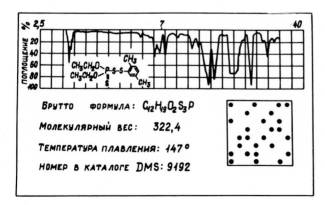

Fig. 11. The output of data from holographic storage.

An alternative solution has been suggested by the Institute of Automatics and Electrometry. It involves formation of a data base, using holographic storage (Figs. 10, 11). A hologram may contain a curve description both in abbreviated digitized and in full form. The creation of optical-mechanic systems for quick survey of holograms, where one cassette drum may involve several hundred plates, each containing hundreds of dozens of holograms, has made it possible to form, on the basis of holographic storage – computer complexes, data banks with data bases of great capacity. This has also enabled storage and visualization if both alphanumeric and graphical information.

The use of computers opens up interesting possibilities for investigating the problems lying at the junction of Chemistry and Biology. Thus, under the guidance of Prof. Vadim A. Ratner, computer modelling of Structure, Function and Evolution of informational macromolecules (proteins and nucleic acids) has been carried out at the Institute of Cytology & Genetics of Novosibirsk Scientific Centre.

```
14                              A
15                              C
16                              C
17                        5'    A
18                              G*C
19                              C*G
20                              G*C
21                              G*U
22                              A*U
23              G               U*A           60
24         U    A               U*A         C U
25         U    C U C G A U   G A C A C      A
26         G    * * * *         * * * * *      G
27         G    G A G C       C U G U G      C
28           G A        G         C       U U
29            20          C*G   A     U
30                        C*G       G G
31                        A*U
32                        G*C 40
33                        A*U
34                      C   A
35                      U   U
36                      G A A
37                    (======)
38                    ANTICODON
 *
```

42   FIG.12.THE COMPUTATED LOWEST ENERGY "CLOVER LEAF" SECONDARY
43   STRUCTURE OF THE YEAST T-RNA PHE. ANTICODON POSITION IS
44   INDICATED.

At present, experimental data are being rapidly accumulated on the protein and nucleic acid sequences and their spatial structure. The analysis of this ample information enables us to achieve a better understanding of spatial self-organization of such macromolecules and of the mechanisms of their function and evolution. The following examples illustrate this.

Based on the polynucleotide sequences and thermodynamic parameters of interactions between nucleotide pairs (via hydrogen bonding, charge transfer and so on) the lowest energy secondary structures of RNA were computed. Fig. 12 presents the results of such computation for the secondary structure of yeast t-RNA^{Phe} This "clover leaf" structure is consistent with the direct X-ray data.

Let us suppose further, that we are interested in the estimation of the consequences of mutational nucleotide substitutions within this macromolecule. (Figs. 13 and 14) show the computed secondary structures of yeast t-RNA^{Phe} in the case of G --→ C (10) and A --→ G (62) substitutions. You can see that in both examples the "clover leaf" structure is disrupted and the function of such macromolecules is supposed to be lost.

```
   45      /



   47                                            A
   48                                            C
   49                          G A G             C
   50                        C       A           A
   51                        G       G          G*C
   52                          C*G              C*G
   53                          C*G              G*C
   54                          A*U              G*U
   55                          A*U              A*U
   56                          C*G              U*A            60
   57        ANTICODON <=====> U*A       10     U*A         C U
   58                   A A G      C U C C A U     G A C A C   A
   59                   U        * * * * *       * * * * *   G
   60                   A U C U G G G A G G U  C   C U G U G   C
   61                        40                               U U
    *
```

```
   65               FIG13.THE COMPUTATED LOWEST ENERGY SECONDARY STRUCTURE OF
   66               THE YEAST T-RNA-PHE WITH SUBSITUTION G-)C (POSITION 10),
   67               DISRUPTING THE "CLOVER LEAF".
    *
```

In the case of yeast t-RNA^{Phe} the computer estimation was made for the whole range of nucleotide substitutions. The results are summarized in (Fig. 15).

Similar   modelling of spatial structure was accomplished for
a series of protein molecules   (for sperm   whale myoglobin as an
example).

The computer   modelling   of substitutions   or   deletions   of
amino acids   in proteins and of nucleotides within nucleic acids
is   of   great   importance   for   modern   Bioorganic   Chemistry,
including Gene Engineering.

```
104                          5'   A G A A U U C G C A C C A
105                              G*C
106                              C*G 62
107                              G*C
108                              G*C 60
109                              A*U
110                      U U U*A G
111                    A             C
112                    G             U
113                      C U C*G U
114                          A*U
115                    U G     G
116                    U G     U
117                      G*C
118                    20 G*C
119                    A G A*U G
120              G                   G
121                    C G C*G A
122                      C*G
123                      A*U
124                      G*C 40
125                      A*U
126                    C     A
127                    U     U
128                     G A A
129                    <=====> ANTICODON
*
```

```
133   FIG.14.THE COMPUTATED LOWEST ENERGY SECONDARY STRUCTURE OF THE
134   YEAST T-RNA PHE WITH SUBSTITUTION A->G (POSITION 62),DISRUPTING
135   THE "CLOVER LEAF".
*
```

The next interesting   trend of the   work   of   Prof. V.   A.
Ratner's Laboratory   involves   reconstruction   of   evolutionary
trees for proteins and   nucleic acids. Computer-aided comparison
of   the primary   sequences   of isofunctional   macromolecules   of
different   biological   species,   enables   their   ranging   by
similarity. The evolutionary tree of such macromolecules   is the
concentrated result   of   this procedure, with the reconstruction

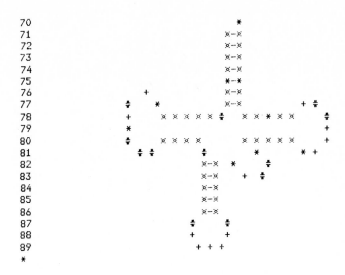

```
70                           *
71                          ×—×
72                          ×—×
73                          ×—×
74                          ×—×
75                          *—*
76                          ×—×
77      ⚏     *             ×—×            +  ⚏
78      +     × × × × × ⚏ × × × * × ×          ⚏
79      *                                     +
80      ⚏   × × × ×         × × × × ×          +
81        ⚏  ⚏     ⚏          *        *  +
82            ×—×  *       ⚏
83            ×—×       +  ⚏
84            ×—×
85            ×—×
86            ×—×
87         ⚏        ⚏
88         +        +
89        +  +  +
*
```

93   FIG.15.THE SENSITIVITY OF THE SECONDARY STRUCTURE OF YEAST
94   T-RNA PHE TO NUCLEOTIDE SUBSTITUTIONS,BASING ON LOWEST
95   ENERGY COMPUTATIONS. DESIGNATIONS: ×— ALL SUBSTITUTIONS
96   DISRUPT THE "CLOVER LEAF" STRUCTURE, *— TWO DISRUPTIVE
97   SUBSTITUTIONS, ⚏— ONE DISRUPTIVE SUBSTITUTION, +— ALL
98   SUBSTITUTIONS ARE NONDISRUPTIVE.
*

of ancestor sequences for each branch point of this tree.

The algorithm has been developed for analysis of evolution of globins and C. cytochromes. Fig. 16 depicts a possible computer   reconstruction of phylogenetic tree   for hemoglobins. This tree was attached to paleontological time scale at 6 points, which had been determined as dated times of existence of common ancestors of some contemporary animal groups. This allowed estimation of the rates of structural changes at different stage of evolution for different parts of macromolecules. It is noteworthy that 500-400 million years ago during the supposed period of Vertebrates coming out to the earth and of appearance of hemoglobin tetrameric structure, the rate was almost 20-times higher than at the next stage. In some contact centres appearing just at this epoch - K_1 $(\alpha_1-\beta_2)$, K_3 (2,3-DPG-binding) - the replacements were fixed at this period only; but later, in some phyletic lines there were no fixed replacements in these centres during 300-400 million years.

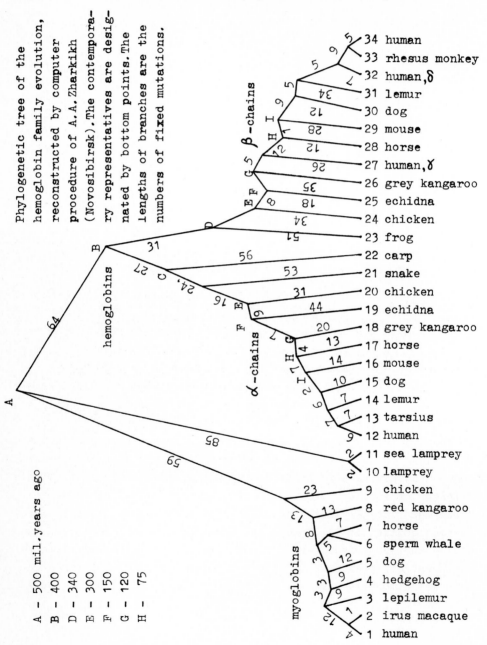

Phylogenetic tree of the hemoglobin family evolution, reconstructed by computer procedure of A.A.Zharkikh (Novosibirsk).The contemporary representatives are designated by bottom points.The lengths of branches are the numbers of fixed mutations.

Fig. 16.

Figure 17. Computerization at Novosibirsk State University

This example illustrates enormous possibilities of computer analysis and modelling for reconstruction of structural evolution of proteins and nucleic acids, and, hence, of the Life on the Earth.

Among other trends of computerized research in chemistry and the related sciences at the Novosibirsk Scientific Centre, it is worth while to mention computer simulation of catalytic processes and reactors developed at the Institute of Catalysis (the head is Academician G. K. Boreskov). This work resulted in modifications of important industrial processes and in new technologies (for example conducting catalyzed reactions in non-stationary conditions). The chemists also widely use computational methods of vibrational spectroscopy, molecular mechanics and quantum chemistry. A central library of programs for quantum-chemical calculations has been created at the Institute of Chemical Kinetics and Combustion, where chemists and biologists may consult specialists on the points of using these programs.

During the last three years, much work has been done on computerization of tuition at the Novosibirsk State University (Fig. 17). Primary stress is laid upon computer modelling of laws and phenomena studied by the students. This considerably facilitates learning and helps students to develop a creative approach to the subject under study. Computerization of tuition at the Novosibirsk State University is being rapidly developed due to a wide use of computerized systems in scientific research developed by the institutes of the Novosibirsk Scientific Centre, including those mentioned in this report.

S.R. Heller and R. Potenzone, Jr. (Editors), *Computer Applications in Chemistry*
© 1983 Elsevier Science Publishers B.V., Amsterdam — Printed in The Netherlands

THE CHEMISTRY IN FUTURE MOLECULAR COMPUTERS

FORREST L. CARTER

Chemistry Division, Naval Research Laboratory, Washington, D.C. 20375

INTRODUCTION

A major current technological U. S. strength today is represented by the semiconducting industry and its innovative, industrious scientists and engineers. Not only is that strength being seriously challenged economically by foreign competitors but it is being challenged by its own success in ever reducing the size of switching elements (transistors) - roughly by an order of magnitude for each ten years. By extrapolation (ref. 1), one may expect that in 20-40 years, the switching elements will be of molecular size, that is in the range of 20 to 100Å. As the size of switches approaches molecular scale, we anticipate two major changes: First, the switch material of choice will no longer be silicon but will be composed of chemical moieties whose structures are natural to switching at the molecular level. Second, the role of the chemist must shift dramatically from that of secondary importance to a role where the creativity and energy of the chemist is a major portion of what promises to be an entirely new industry. It is for these two reasons that I am delighted to be discussing the challenge presented by the chemistry in future molecular computers with educators expert in both chemistry and computers.

The concept of molecular switching in electronic devices became viable in 1978 at the Naval Research Laboratory when it was recognized that communication with the switching moieties was possible by using molecular "wires" of linear polymeric conductors like doped polyacetylene $(CH)_x$ and polysulfur nitride $(SN)_x$ (refs. 1, 2). The enormous implications inherent in the concept resulted in a workshop on the subject in 1981 (ref. 3), and it is reviewed in part here. Earlier in 1974, Aviram and Ratner had discussed a

"molecular   rectifier"   but   one   without   "wires"   (ref.   4).
Elaboration   of   concept   of   molecular   switching,   as   presented
recently   in   Europe   (refs.   5,   6)   and   developed   further   at   NRL,   is
included   in   this   discussion.

The   two   following   sections   treat   the   fabrication   of   a
Molecular   Electronic   Device   (MED)   computer   in   broad   outlines   (ref.
1)   and   the   important   problem   of   bridging   the   dimension   gap   between
the   macroscopic   and   "molecular"   world   (ref.   6).   Surface
modification   chemistry   is   seen   to   be   an   important   contribution   to
both   of   these   areas.   Electron   tunnelling   in   short   periodic   arrays
is   next   discussed   in   order   to   give   an   example   of   one   important
kind   of   molecular   switching.   Having   thereby   demonstrated   the   level
of   complexity   which   might   occur   in   MED   technology,   the   importance
of   understanding   the   power   of   self-organization   as   an   assembly
tool   for   fabrication   is   stressed.   While   conformational   changes
play   a   role   in   electron   tunnel   switches,   they   become   all-important
in   switching   involving   soliton   phenomenon   and   in   "smart"   molecules
that   can   perform   logical   operations.   The   use   of   solitons   in   group
operations,   controlling   tunnel   switches,   mechanical   actuation,
micro-laser   output,   and   amplification   are   then   considered   in
order.   Finally,   MED   memory   units   and   a   look   at   the   future   in   MED
synthetic   techniques   is   considered.

THE SYNTHESIS OF A COMPUTER

Modern   silicon   chips   are   now   so   complex   compared   to   most
chemical   operations   that   it   is   difficult   to   imagine   a   computer
dependent   on   molecular   conformational   switching   without   a   general
description   of   its   fabrication   and   a   technique   for   addressing
molecular-sized   switching   moieties.   As   a   modest   goal   to   keep   in
mind,   one   may   consider   a   three-dimensional   architecture   in   which   a
switch,   its   "wires,"   and   insulation   occupy   a   cube $1000 \overset{\circ}{A}$ on   an
edge.   This   results   in   a   density   of 10^{15} switches/cm^3 (ref.   1).

In   general,   the   microfabrication   of   such   a   three-dimensional
computer   might   be   accomplished   by   three   methods:
   (a)   Lithographic   and   vapor   deposition   techniques.
   (b)   Chemical   synthesis   involving
      (1)   Merrifield   techniques
      (2)   Surface   modification   techniques
   (c)   Biotechnical   synthesis   of   computer   ultracircuits   and
      assembly.

The first approach represents an extension of current semiconductor technology and was briefly discussed earlier (ref. 1, p. 144). The third approach was suggested by K. Ulmer at the recent MED Workshop (ref. 7) and presumes that recombinant DNA techniques and genetic engineering can achieve the fabrication of self-assembling "molecular" electronic components.

The philosophy of this paper, however, is in greater sympathy with a more synthetic chemical method of computer fabrication. In this second hypothesized approach to manufacturing a three-dimensional array of molecular switches, i.e., a computer, considerable use is to be made of solution reactions, sometimes catalyzed in selected regions near the growing surface by a laser beam of selected wavelengths.

To appreciate and imagine how a computer might be synthesized or grown in solution, it will be useful to acquaint the reader with the process known as the Merrifield Method of polypeptide synthesis (ref. 8). This highly successful technique of biochemistry is now a standard method for producing long polypeptide chains in high yield. Furthermore, these polypeptide chains, perhaps a hundred amino acid units long, might be composed of many different amino acids, linked together in a very specific and preselected order. The process of building the polypeptide chain is initiated by preparing a substrate, usually in the form of plastic beads, with a link to the first amino acid that can later be broken. Then, using a system of automatic controlled valves, various reagents and solutions containing the appropriate amino acids are reacted with the polypeptide growing on the bead substrate. When the desired polypeptide is achieved, the linkage to the substrate is broken and the polypeptide collected.

The proposed process for growing an array of molecular devices is similar, except that in different areas of the now planar substrate, one is producing regions of different character, i.e., conductors, insulating material, and components which will be active molecular devices. Different solutions will be introduced, reacted, and rinsed away for each layer of each function.

A schematic overview of the process was suggested in (ref. 1), Fig. 21. By employing optical lithography, large electrical contacts to the external environment would be developed on a substrate of high thermal conductivity. Then, by using the technique of forming valence bonds between a molecule and a metal surface, described in the next section, specific areas would be covered with molecules destined to become conducting filaments and paths, while the remaining surface is covered with insulating material. Each new layer contains reactive sites for chemically bonding the succeeding layers. The fine graphic detail necessary here might be achieved by using x-ray lithography or 'molecular' lithography like epitaxial superstructuring discussed recently (ref. 6). At this point, the prepared substrate would be introduced into "reaction cell," into which the flow of reagents, solutions, and rinses is controlled by a computer. Thus, layer by layer, active elements, conducting filaments, and insulating materials are built up, just as a long chain polypeptide is built up in the Merrifield method. In some instances, the reactions could be catalyzed in certain regions by a "reaction laser". Ultimately, additional electrical contacts to the external environment would be developed on the outer surfaces.

Such a process presumes, of course, that the chemistry of each step is well understood and controlled. Incomplete reactions are difficult to tolerate in such a case. Electrochemical cell techniques might be employed to both drive reactions to completion and to control directional assembly of components. In addition, enough duplication must be built in so that there is a finite overlap of functions. Thus, a conducting filament might involve several strands of $(SN)_x$ molecules.

The resulting envisioned chemical computer of the future might have a central processor unit and main memory with a volume of a cubic centimeter. It would probably be mounted on a support cooled by a small helium refrigerator. The electrical leads would be relatively few, with most of the input timing control pulses and numerical input data being transmitted via optical pulses through cooled masks. Output is envisioned as being via multichannel optical techniques in which each channel source would be highly directional. A "molecular" microlaser for achieving such an output has been proposed earlier (ref. 1, p. 147), and is elaborated further below in the Molecular Microlaser Section.

MOLECULAR COMMUNICATION AND SURFACE MODIFICATION CHEMISTRY

Both polyacetylene, $(CH)_x$, and polysulfur nitride, $(SN)_x$, are to first order one-dimensional conductors (refs. 1, 9). The conductivity of sheet $(CH)_x$ can be changed twelve orders of

magnitude by doping to that of a poor metal, while $(SN)_x$ is
normally a metallic conductor.

If polyacetylene, $(CH)_x$, and sulfur nitride, $(SN)_x$, can
perform as "wires" to the "molecular" switches, one still requires
a communication transition from the macro-world to the "molecular"
world of a single or a few chains of $(SN)_x$. We have suggested that
the preferred method of communicating electrically with a
molecular switch was via a direct Valence-Bonded Surface (VBS)
contact (ref. 1, p. 129). Here, a conducting polymer is covalently
bonded to a metal surface (like Pt) by first exposing the metal
surface to a reactive trimethoxysilane and then chemically
modifying the attached silane to form:

$$\text{metal-Pt-O-Si}(OCH_3)_2CH = CH\text{-}CH = N\text{-}(SN)_x$$

This approach, typical of current surface modification chemistry,
however, results in a short insulating section between
metallically conducting $(SN)_x$ and the Pt metal. This potential
problem would appear to be avoidable by noting the recent
preparation of the sulfur diimides by Scherer and Wies (ref. 10)
and of $(SN)_x$ from species I by Milliken (ref. 11).

$$(CH_3)_3Si-N{=}S{=}N-Si(CH_3)_3$$

I

II

Since in both sulfur diimides, nitrogen is bonded to
tetravalent silicon, we can reasonably anticipate that the diimide
could be bonded directly to the clean surface of semiconducting
silicon as follows:

$$\text{semiconducting-Si-N} = S = N\text{-Si}(CH_3)_3$$

although a sulfur link to the surface might be stronger. Then for
successive treatments of the modified Si surface by first SCl_2 and
then the sulfur diimide I, one can increase the size of the chain
each time by exactly one - $(SN)_2^-$ increment. Whereas Milliken

produced (SN)$_x$ by the sublimation of the black powder reaction product of SCl_2 and $((CH_3)_3SiN)_2S$ (I), the above approach permits the control of the length of the (SN)$_x$ chain by units of $-(SN)_2-$.

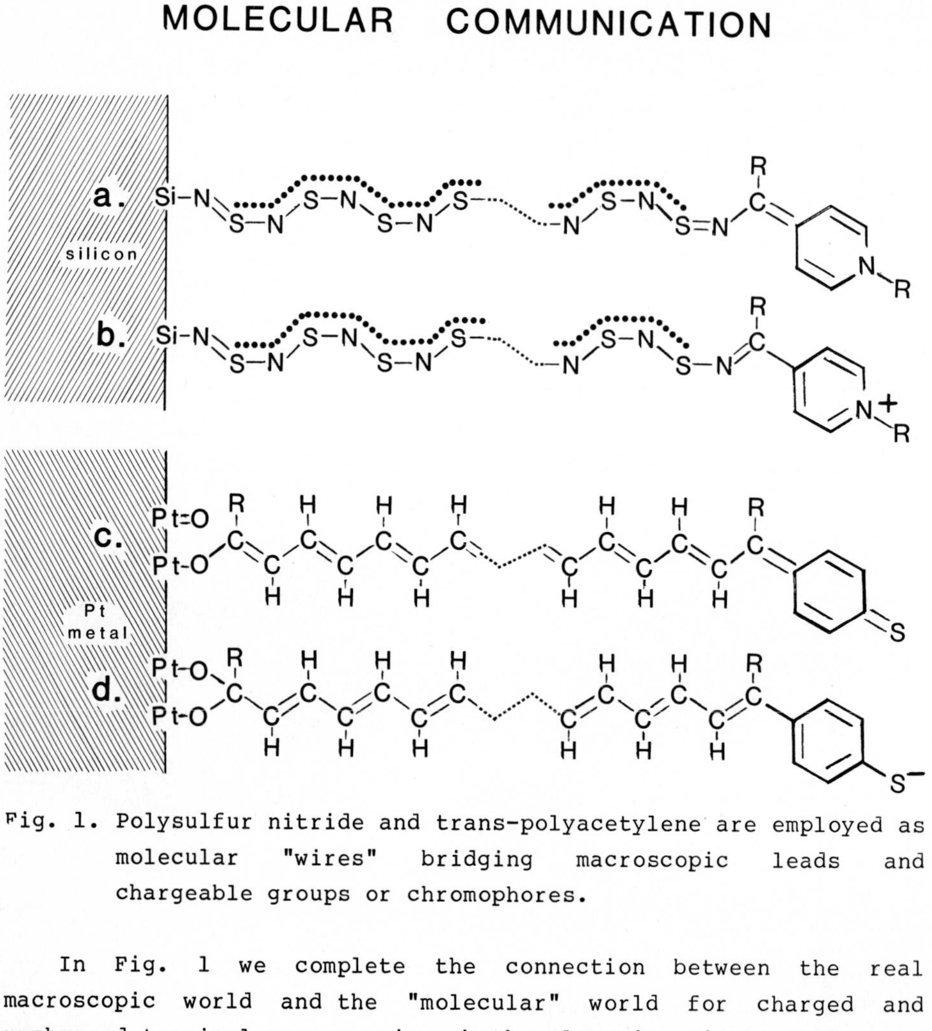

MOLECULAR   COMMUNICATION

Fig. 1. Polysulfur nitride and trans-polyacetylene are employed as molecular "wires" bridging macroscopic leads and chargeable groups or chromophores.

In Fig. 1 we complete the connection between the real macroscopic world and the "molecular" world for charged and uncharged terminal groups using both polymeric single chains of (SN)$_x$ and (CH)$_x$. In Fig. 1a and 1b, we note that only the conformational change at the terminal group is readily discernible, since (SN)$_x$ (a poor metal) has a band-like conduction mechanism involving a delocalized antibonding π^* orbital. However, in Fig. 1c and 1d, the conformation changes in both the terminal group and (CH)$_x$ chain are depicted. Figure 1 constitutes a plausibility argument since such systems have not in fact been demonstrated at this time. However, the fabrication of such links from the macroscopic world to the molecular world seems to be achievable through further development of the techniques of surface modification chemistry (review, ref. 12).

The importance of surface modification chemistry to MED technology is immediately recognized when it is understood that the proposed synthesis of a MED computer via a Merrifield-type technique consists of a long series of surface modification reactions. Thus, most all reactions involve one reagent whose movement is restrained and a second whose direction of attack is limited by the plane of the substrate and other attached groups. Added to these conditions is the need for the attached reaction product to fold into the growing substrate in a spacefilling and functional manner. The theoretical and computational challenges presented by the application of surface modification chemistry to future MED technology should make for a fertile and rewarding new field.

ELECTRON TUNNEL DEVICES

Among the most promising of switching mechanisms tabulated earlier (ref. 1) is that of electron tunnelling in short periodic structures. This switching mechanism is based on the quasiclassical approach of Pschenichnov (ref. 13). The concept is illustrated in Fig. 2, where potential wells and barriers are arranged in a short periodic array. Between the barriers, pseudo-stationery energy states are represented by dotted lines in Fig. 2a. If the energy of an approaching electron (from the left) matches a pseudo-stationery level, Pschenichnov indicated that the electrons are transmitted as if the barriers were absent (Fig. 2b). Recently, the mathematics of tunnelling through an arbitrary set and number of step function potentials has been derived using standard quantum mechanical methods (ref. 14). These exact results confirm those of Pschenichnov.

The switching action will occur if the barrier heights or wells can be controlled externally by the motion of a positive or negative charge, thereby switching off the electron tunnelling flow by shifting some of the pseudo energy levels. Using this concept, molecular analogues of semiconducting NAND and NOR gates have been proposed in 1979 (ref. 1, Fig. 17).

TUNNELLING PERIODIC BARRIERS

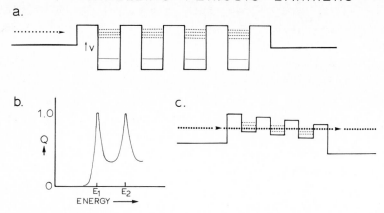

Fig. 2. In part a the transmission of the electron approaching from the left through the periodic barriers is perfect if its energy matches exactly the pseudo-stationary energy levels E_i (dotted lines). Its transmittance $Q(E)$ is schematically indicated in part b.

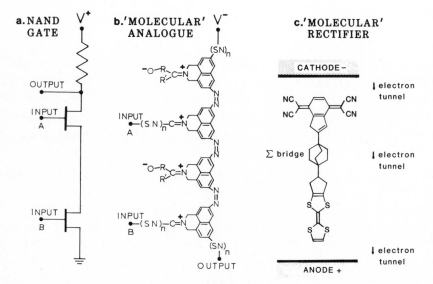

Fig. 3. Early molecular analogues of NAND gate, parts a and b, and the molecular rectifier of Aviram (4), part c, are illustrated. While the molecular analogue of a NAND gate, part b, is based on a single tunnelling process involving a periodic barrier (1) the molecular rectifier of Aviram and Ratner (4) involves three independent tunnelling processes.

The "molecular" NAND gate is illustrated in Fig. 3 as part b, while its semiconducting analogue is part a. The periodic potential is established by the four quaternary nitrogens located on the four diazo-linked napthalene derivatives. The system we describe as four <u>Control</u> <u>Groups</u> (CG) on the switch <u>Body</u>. Two of the Control Groups can be controlled by charge flow down the $(SN)_x$ chain, thereby neutralizing the quaternary nitrogen plus charges. The other two are dummy Control Groups.

For comparison we also show in Fig. 3c the earlier suggested "molecular" rectifier of Aviram and Ratner (Ref. 4, 1974). They proposed that current would tunnel from the cathode to the acceptor to the donor through the insulating bridge, and finally from the donor to the anode, but not in the reverse (here, up) direction. The acceptor portion of the molecule is related to the well known TCNQ (tetracyanoquinodimethane) species, while the donor is modelled after TTF (tetrathiofulvalene). By contrast with the above NAND gate, the three tunnelling processes are independent and a technique for communication with the rectifier has not been detailed. Surface modification techniques might be useful in this regard, or Langmuir-Blodgett film techniques, as has been proposed by Aviram (ref. 15).

More recently a molecular analogue to the "NOR" gate of conventional electronic circuitry was suggested (ref. 14) and is illustrated in Fig. 4. The molecular switch of Fig. 4a is the analogue to the NOR gate of Fig. 4b. Its Body is based on a fluorine bridged stack of gallium phthalocyanine rings similar to the compounds recently reported by Kuznesof, Wynne, Nohr, and Kenney (ref. 16). The stack is composed of rings of Type C (Fig. 4c) where the $-(SN)_n-$ linkage is replaced by the $-CH_2SO_3-$ groups: these latter are attached to the imino dummy Control Groups which are necessary to develop the short periodic potential. The D rings (Fig. 4d), nickel phthalocyanine moieties, provide links to both the ground and the negative potential through the Ni-S bonds (see Fig. 4a), as well as to the $(SN)_n-$ output lead. These terminating rings also serve to provide a suitable environment for the stacked bridged gallium phthalocyanine Body.

The enormous flexibility available in molecular electron tunnel switches is hardly apparent in the examples given to date in Figs. 3 and 4. However, earlier (ref. 2), we had pointed out that at least four classes of Control Groups can be readily

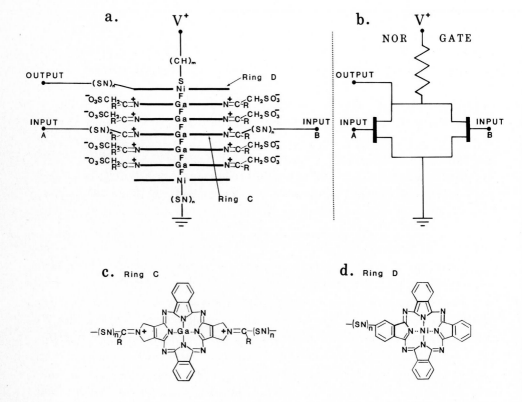

Fig. 4. A molecular analogue to a NOR gate based on a stacked
fluorine bridged gallium phthalocyanine-type ring. The
bridging fluorines result in an insulating barrier between
the rings while the terminal sulfurs provide a strong
multivalent bond to the $(SN)_n$ and more resistive $(CH)_n$
conductors.

identified. These are: (1) Charge Flow Control Groups, such as
appears in Fig. 3 and the input leads of Fig. 4a, where a charge
is changed, induced, or neutralized by charge movement along a
$(SN)_x$ or $(CH)_x$ conductor; (a) Dipole Control Groups, where a
change in dipole direction is induced by an electric field or
photon absorption. An example exists in the enol-keto
tautomerization; (3) Photochromic Control Groups, where a very
large change in an electric dipole field is induced in a
zwitterionic or photochromic molecule; finally, (3) Binuclear
Organo-Metallic Control Groups in which reciprocal changes in
transition metal charges and valences are induced by photon
absorption and/or an electric field.

These four classes of Control Groups provide the future device
fabricator an unparalleled opportunity to build in, via a choice
of molecular structure and composition, Control Groups with a wide
variety of properties. These properties can include the chemical
or structural equivalent of an electric bias, as well as a
variation of relaxation time constant of at least ten orders of
magnitude.

The complexity of the switches in Figs. 3 and 4 represent
significant synthetic challenges; however that is only part of the
problem. Structural conformation and packing are clearly important
considerations. The next section may serve as a useful signpost in
that regard.

SELF-ORGANIZATION

The chemical synthesis of a complex MED device or processor
via a Merrifield-type process, as described earlier in the
Synthesis of a Computer Section, will not be possible unless the
entire synthetic process and organization is outlined in advance
in minute detail. Currently, this is not feasible. Not only must
each reaction go to completion but the product must organize
itself so that it steers the course of the subsequent reactions
and then packs itself in a space-filling manner against the
growing substrate. To prescribe such a course of events is far
beyond our current computation capabilities; for example, only
modest success is achieved in predicting conformations in short
polypeptide chains (ref. 7, see Discussion). Nevertheless, an
interesting and useful example of the possibilities is suggested
by the article of F. H. Stillinger and Z. Wasserman, entitled

"Molecular Recognition and Self-Organization in Fluorinated Hydrocarbons" (ref. 17). The important principle that can be extrapolated from their work is that the various chemical components should be so designed and assembled that the computer or device is largely self-organizing.

CONJUGATE PAIRING and SELF-ASSEMBLY

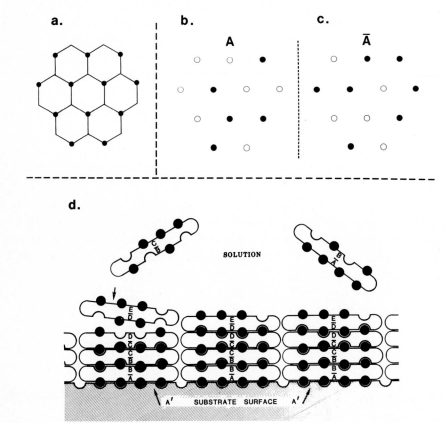

Fig. 5. In perhydrocoronene twelve C-H bonds project up as in the
         dotted carbons in part a, twelve project down and twelve
         are in the plane. Part b indicates a possible replacement
         pattern of five fluorines for hydrogen that project up,
         while part c is the corresponding conjugate (17). Part d
         suggests the use of the attractive potential between
         conjugates to produce an insulating layer exactly four
         molecular layers thick.

These authors demonstrated the self-organizing concept by looking at the pair potentials for a family of saturated partially fluorinated fused-ring hydrocarbons. For example, in Fig. 5a we indicate schematically perhydrocoronene, which has twelve hydrogens on each side which project normal to the general plane of the molecule. For each side there are a total of 1376 different patterns that can result if these twelve hydrogens are replaced by fluorine (i.e., by 0 to 12 F's). One such possibility is indicated in Fig. 5b, while Fig. 5c shows what Stillinger and Wasserman call its "conjugate." If the pattern of Fig. 5b is indicated by A and is the top of one fluorinated perhydrocoronene (FPHC), and Fig. 5c is symbolized by $\overline{A}$ and is the bottom of another FPHC, then the two can fit together perfectly. Such a pair has a deep well in terms of its pair potential, while face A with any other FPHC has a very shallow pair potential curve, if a minimum even exists. Now, since each FPHC has two sides, one could develop a small crystallized stack or series like (F/A) ($\overline{A}$/B) ($\overline{B}$/C) ($\overline{C}$/D), etc., where each face, A, B, or C is attracted only by its conjugate face, $\overline{A}$, $\overline{B}$, or $\overline{C}$.

By using these concepts of self-organization, we will now indicate how a prescribed thickness of insulation can be deposited or a definite length of insulation can be obtained from a single solution (ref. 2).

Consider a substrate surface (either organic or inorganic) that is very similar to A, call it A', and hence is a surface on which $\overline{A}$ would be strongly expitaxially attracted, as in Fig. 5c. Then by using a _single_ solution containing the mixed FPHC species ($\overline{A}$/B), ($\overline{B}$/C), ($\overline{C}$/D, and ($\overline{D}$/E), one could put down an insulating layer on the A' substrate which in this case is _exactly_ four FPHC molecules thick (and no more) and having a top surface specified by E. Such a solution as we have just described might be termed a "pre-organized" solution.

This possibility of engineering with relative weak pair potentials while paying close attention to packing considerations provides a useful step of fabrication beyond the Merrifield approach of molecule-by-molecule buildup and begins to approach what must happen in some biological systems.

By linking pairs of FPHC molecules together with an oxymethylene bridge, the possibility of making microtubules was described by Stillinger and Wasserman (ref. 17) and is indicated

in their Fig. 17 (or ref. 2, Fig. 6). This technique is, of
course, an interesting method of providing molecular conducting
filaments, like strands of $(SN)_x$, with an insulating sheath. By
preparing a pre-organized solution of bridged FPHC pairs, the
length of the "molecular spaghetti" could also be determined in
advance (ref. 2).

SOLITON SWITCHES AND CONFORMATIONAL CHANGES

Both electron tunnel switches and soliton switches involve
conformational changes at the molecular level, however, in soliton
switches they constitute the major effect and hence may be of more
interest to chemists. Now several kinds of solitons exist; for
example, in the biological context a soliton or "solitary wave"
has been proposed as the mode of energy transport down the a-helix
by Davydov (ref. 18). We shall be more interested in a second kind
of molecular soliton, one which has been proposed as the principal
electrical conduction mechanism in weakly doped transpolyacetylene
$(CH)_x$ (refs. 9, 19). In the first example, the soliton is
primarily a vibratory exciton strongly associated with the dipole
moment of the polypeptide amide bond, while in the second example,
it is a charged soliton associated with a hole or electron in a
p-orbital of carbon. Radical solitons in $(CH)_x$ have also been
strongly implicated in the conduction mechanism by Kivelson
(ref.20). In essence, a soliton is a pseudo particle of definite
energy, momentum, and velocity, whose transport involves no
dissipation of energy and whose associated differential equation
always involves a non-linear coupling.

For the purposes of this discussion, the important point is to
note that the passage of a soliton interchanges single and double
bonds in trans-polyacetylene. This is illustrated in Fig. 6a,
where the soliton is moving from left to right. The length of the
soliton is appreciably larger than illustrated and probably
extends over 10 to 14 carbon atoms.

Figure 6 also illustrates a valence bond representation of the
newly suggested mechanism of polaron transport in conjugated
systems proposed by Campbell and Bishop (ref. 21). While a soliton
can propagate in trans-polyacetylene with a conformation change
(Fig. 6a), it cannot in cis-polyacetylene because the resultant
conformation change is associated with a non-zero energy change
per unit length. However, a polaron can propagate in cis-$(CH)_x$

because there is no net conformation change (Fig. 6d and e).

SOLITONS AND POLARONS

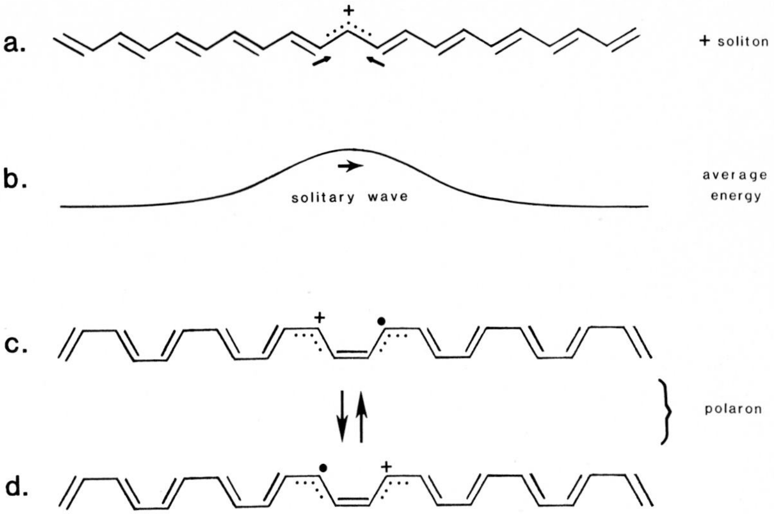

Fig. 6. While the motion of a soliton in transpolyacetylene is
associated with a moving "phase boundary" or "kink" (part
a); the motion of a polaron produces no net configuration
change (see part c,d). Solitons will not propagate in
cis-polyactylene but polarons can.

However, since a molecular switch depends upon conformational
changes (ref. 22), we need only consider soliton propagation here.
To see how this might work, let us incorporate a chromophore
perpendicular to the (CH)$_x$ chain so that it shares at least two
carbon atoms. As the chromophore consider the "push-pull" olefin,
1,1-N,N-dimethyl1-2-nitroethanamine, in its ground and activated
states as illustrated below:

PUSH-PULL   OLEFIN

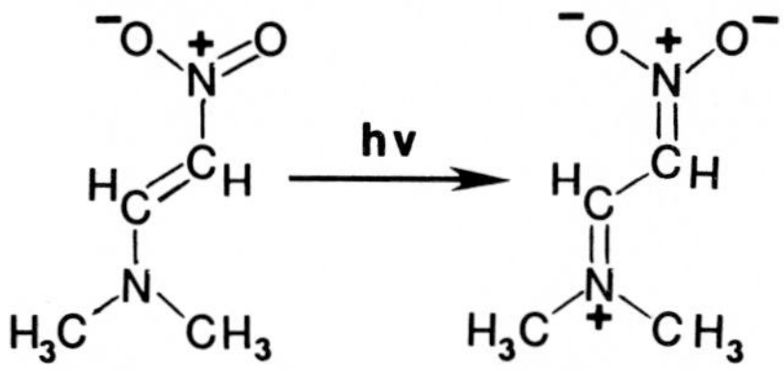

SOLITON   SWITCHING

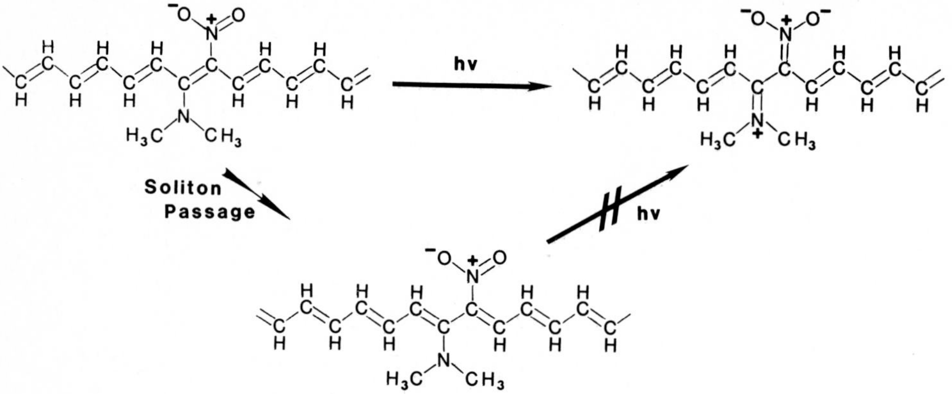

Fig.   7a.   A   push-pull   disubstituted   olefin   imbedded   in
        transpolyacetylene   can   1)   be   switched   off   by   the
        propagation   of   a   soliton,   or   2)   be   used   as   a   soliton
        detector.

SOLITON   SWITCHING

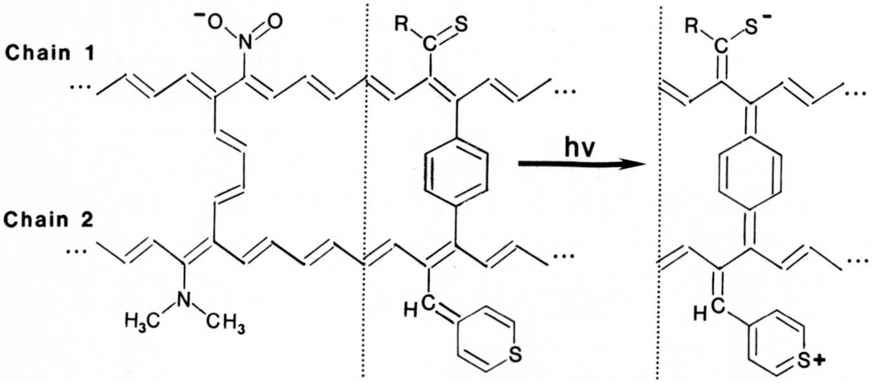

Fig.   7b.   Soliton   switching   involving   two   trans-polyacetylene
        chains   and   two   chromophores.   The   second   sulfur
        chromophore   can   be   photoactivated   but   not   the   first
        chromophore.

Note that the double bond shifts away from the central carbon atoms. In Fig. 7a, these carbons have been incorporated as part of a larger $(CH)_x$ chain. In the ground state, the push-pull olefin (Fig. 7a, left) should still undergo photoactivation using polarized light. However, if a soliton has been propagated down the polyene chain, then that photoactivation process can no longer take place (see Fig. 7a, bottom). Thus the soliton has switched off the internal charge transfer reaction. (Note also that the absorption spectra of the push-pull olefin can serve as a detector for the passage of the soliton.)

The soliton switching concept can be extended to two chains and two different push-pull structures of extended chromophores. This is indicated in Fig. 7b, where the conformation of chain 1 has switched off the nitro-amine chromophore but does not prevent the photoactivation of the sulfur-containing chromophore. The passage of a soliton down chain 1 will turn the first chromophore on and the second off; a soliton moving down chain 2 will turn both of them off. In Fig. 8, the concept of soliton gang switching is generalized where A, C, and D are generalized electron acceptors, conjugated connectors, and electron donors, respectively. Notice that each chromophore, separated from others by dotted lines, has a different relationship to the conformations of chains 1, 2, and 3. This relationship is summarized in Table 1 where eight different chromophore-chain relationships are indicated as possible. Only the first four relationships indicated in Table 1 are illustrated in Fig. 8. For reasons to be offered shortly, the chromophores are identified as channels in Table 1. In Fig. 8, note that the relationship of the firstmost chromophore to each chain is such that the chromophore is subject to photoactivation; in Table 1 this set of relationships is indicated by a vertical row of 1's. However, the second chromophore or channel is turned off by the first chain, while the third channel is turned off by the second chain. In short, the chromophores or channels are so arranged that the passage of a soliton down any of the three chains turns off any chromophore that is on and turns on one of the others. Table 1 then expresses the demultiplexing concept that soliton propagation can be used to perform gang switching where three input states can control eight different output channels. By separating the chromophores by about 200 Å the density of such switches can be estimated to be as high as

10^{18}/cc! It should be remembered however, that soliton propagation is less than the speed of sound.

Table 1. Three Chain Gang Switching

Channel[a]	1	2	3	4	5	6	7	8
Chain[b] 1	1	0	1	1	1	0	0	0
Chain 2	1	1	0	1	0	1	0	0
Chain 3	1	1	1	0	0	0	1	0

[a]A channel is only open if a vertical column is all ones.
[b]Soliton passage through a chain changes all ones to zeros and vice versa

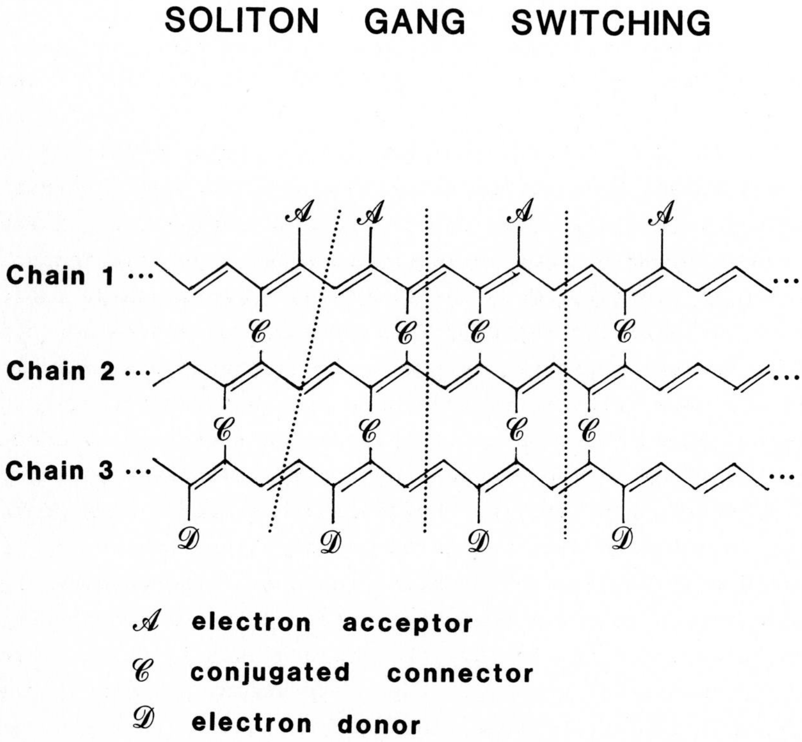
SOLITON GANG SWITCHING
Chain 1
Chain 2
Chain 3
𝒜 electron acceptor
𝒞 conjugated connector
𝒟 electron donor

Fig. 8. Soliton gang switch showing four of eight possible different arrangements chromophores can have relative to three conjugated chains.

SOLITON GENERATION AND SOLITON REVERSAL

While solitons might be generated by photoabsorption, such a technique does not have the spatial resolution desired for control of molecular electronic devices. An alternative method for the generation of solitons which is more amenable to MED technology has been suggested (ref. 22). This technique makes use of proton tunnelling in an electric field E, as indicated below:

PROTON TUNNELLING & SOLITON GENERATION

As the proton, H^+, moves to the right to form a new bond with nitrogen in the presence of the electric field, the valence electrons (usually π electrons) move to the left, as is suggested by the curved arrows. After the conformational changes occur, as indicated in the lower equation, we can imagine a trivalent carbon anion being formed on the left and a carbonium ion being formed on the right. Further motions of these charged states to the right and left, respectively, are suggested by the wavy arrows in the lower equation above. However, since the proton tunnelling is necessarily associated with conformational changes induced by an electric field, it seems that the necessary elements for charged soliton formation are present. Accordingly we anticipate that a negative soliton will move to the left, while a positive soliton will propagate to the right.

At this point (lower equation), the system has generated two solitons of opposite charge moving in opposite directions, but then soliton generation stops. If the potential is increased sufficiently, then a hydride ion (H^-) can leave the right hand nitrogen, hopping back to the oxygen, and generating two more solitons of the same charges and moving in the same direction as before. Moreover, when the solitons depart and their potential has dropped sufficiently for the hydrogen to see primarily the electric field E, then proton tunnelling will occur again and generate two more solitons. In short, we will have an oscillatory generation of solitons. Finally we note that by changing the end groups, proton and hydride ion tunnelling will occur at a different potential. In short, one has a structural control over an effective built-in electrical bias.

It has been proposed that the direction of soliton propagation can be reversed via the use of bond conjugation within ring systems (ref. 22). Via this technique, a soliton can be returned on either a single $(CH)_x$ chain or a second parallel $(CH)_x$ chain.

'SMART' MOLECULE

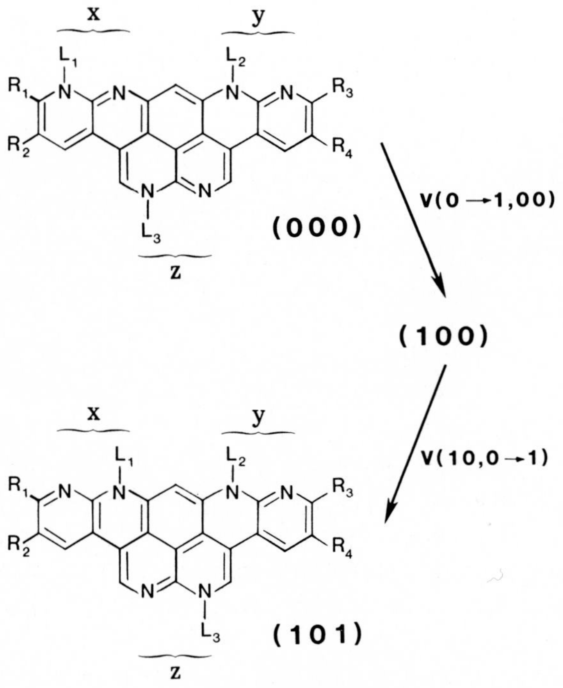

Fig. 9. Two of the eight possible ligand distributions are illustrated here along with the single ligand shift frequencies responsible for one conversion path between them.

"SMART" MOLECULES

An interesting and potentially very useful concept was introduced at the MED Workshop by R. C. Haddon and F. H. Stillinger (ref. 23) under the designation of "smart" molecules. The more sophisticated version of such molecules which they discussed is indicated in Fig. 9. Information storage in such a molecule is possible by a movement of the labeled groups, L_1, L_2, L_3, to the adjacent nitrogens. These groups, L_i, might be H-, CH_3-, or $(CH_3)_3Si$-; however, the heavier groups are preferable in order to have high barrier heights and hence preclude the inadvertent tunnelling and thermal transfers which would destroy information.

If the eight different tautomeric states of the molecule are of different energy and all the switching energies (to shift the L_i) are coupled and non-equivalent, then information can be processed within the molecule (ref. 23). In order to achieve this optically, Haddon and Stillinger required twenty-four different non-overlappinng frequencies, i.e., three for each tautomer state. Hence, for the (010) state, the three switching frequencies would be designated as $v(0 \to 1,1,0)$, $v(0,1 \to 0,0)$, and $v(0,1,0 \to 1)$. By using one bit of the three in a "carry"-like capacity, rudimentary binary operations between the other two bits can be obtained by the appropriate sequence of optical sequences. For example, a five pulse sequence simultaneously yields the binary sum (mod 2) and the product or logical "and." Other optical pulse sequences produce the logical "or", "nor", and "nand" results.

Although communication with their "smart" molecule was achieved via optical means, thus avoiding the question of the addressability of molecules (moieties), the conceptual exercise of Haddon and Stillinger is quite important in that it illustrates the importance of conformational changes as logical operations and points the way to other "smart" moieties. In the next section rather different logical operations are achievable at the molecular level, using solitons as the signal and configurations of soliton valves as the device.

THE SPECIAL PROPERTIES OF SOLITON VALVE CONFIGURATIONS

Conformational changes are especially interesting in configurations of soliton valves. A soliton valve (ref. 22) and its properties are suggested in Fig. 10, where we see three

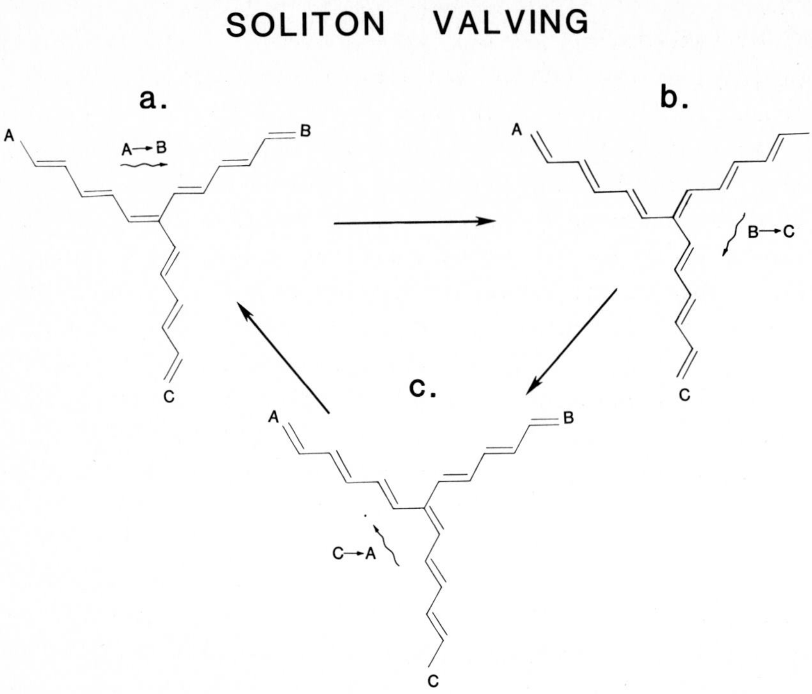

SOLITON   VALVING

Fig. 10. The propagation of a soliton from A to B corresponds to a
         clockwise   rotation   by   120°   of   the   upper   left-hand
         configuration. This valvelike action is   related   to   but
         not isomorphous with a threeway valve.

TRINUCLEAR
CYANINE DYE

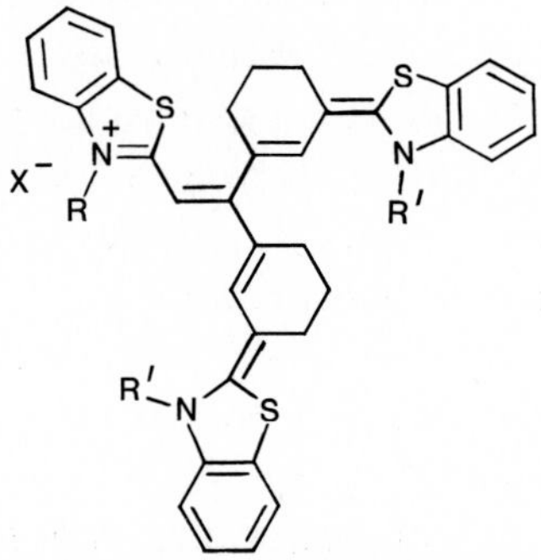

Fig. 11. Cyanine dyes   are well known sensitizers for photographic
         emulsions; the   trinuclear dye   shown here   has   the same
         conjugation configuration as the valve of Fig. 10a.

conjugated half-chains joining at a single carbon (branch atom).
The propagation of a soliton through the branch carbon is
comparable to a threefold rotation operation. Thus a soliton
moving from A to B (or B to A) in Fig. 10a switches the double
bond at the branch carbon from the A chain to the B chain. Note
also that the two adjacent single bonds on the path from B to C
(or C to B) preclude that being a soliton path in Fig. 10a. Thus a
soliton valve is suggestive of a three-way valve, where each
moving soliton, in addition to its transport of energy and/or
charge, corresponds to the right or left hand rotation of the
valve. Hence the passage of one soliton strongly influences the
path of the succeeding soliton. As an example of a molecule having
a conjugated system like a valve, consider the tri-nuclear cyanide
dye of Fig. 11 (ref. 24).

THREE-VALVE CYCLIC CONFIGURATION

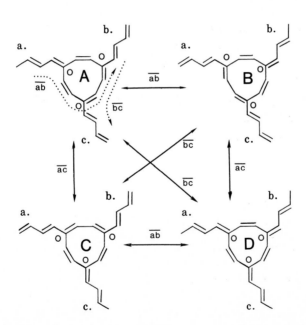

Fig. 12. Three valves in a cyclic arrangement, each separated by
         three bonds, have the four possible configurations shown
         (A--D). Soliton propagations, like ab, from chain a to
         chain b, are operations that connect one configuration to
         another. In this case, these operations (soliton
         propagations) form a group.

From a paper chemistry point of view (i.e., ignoring stereo-chemical considerations), arrays or configurations of such valves can be arranged in either linear or cyclic configurations. In Fig. 12, we illustrate the four possible conjugated bonding configuratios, A → D, in a cyclic configuration having a single nine carbon atom ring. The "O" at each branch carbon indicates the location of two adjacent single bonds and, hence, an "occluded" soliton turn or path. The operation associated with the passage of a soliton from the end of chain a to the end of chain b is designated as $\overline{ab}$. In the systems of Fig. 12, the pathway is usually indirect and includes changing the configuration at all three branch atoms, as indicated by the dotted lines in Fig. 12. Note also that $\overline{ab} = \overline{ba}$, etc. The three operations, $\overline{ab}$, $\overline{ac}$, and $\overline{bc}$, with the inclusion of the identity operation, E, form a group, as indicated in Table 2. It is also very interesting that, while these operations are represented in a plane in Fig. 12, the group (Table 2) is isomorphous with the three-dimension group, D_2, which corresponds to three perpendicular twofold axes. The above results clearly demonstrate that soliton propagation in conjugated systems can correspond to sophisticated logical operations.

Table 2. Group Table for Soliton Transformations in a
         Three-Branch Single Cycle of Valves with
         Odd, Odd, Odd Conjugation.

	E	$\overline{ab}$	$\overline{ac}$	$\overline{bc}$
E	E	$\overline{ab}$	$\overline{ac}$	$\overline{bc}$
$\overline{ab}$	$\overline{ab}$	E	$\overline{bc}$	$\overline{ac}$
$\overline{ac}$	$\overline{ac}$	$\overline{bc}$	E	$\overline{ab}$
$\overline{bc}$	$\overline{bc}$	$\overline{ac}$	$\overline{ab}$	E

E = identity operation; $\overline{ab} = \overline{ba}$, etc.

Although the group operations of Table 2 can be easily shown applicable to all single cyclic conjugated systems with an odd number of bonds between each of the three-branch atoms, other cyclic systems with even number of bonds between the three branch atoms are primarily soliton reversers; those having two sides with even number of bonds and one side with odd number of bonds are operationally the same as an isolated valve. Finally, a single

cycle having two sides with an odd number of bonds and the third
side with an even number of bonds has only four configurations
like Fig. 12, but it neither forms a group nor are all the
transformations physically realizable by soliton passage. It can,
however, function as a switch.

SOLITON CONTROL OF TUNNEL SWITCHES

The next five sections will present various soliton-related
phenomena with heuristic applications to future MED technology.
The generation of pairs of oppositely charged solitons via proton
tunnelling provides another non-optical technique of controlling
electron tunnel switches via Control Groups. This is indicated in
Fig. 13 where a pair of solitons are employed to add opposite
charges to transversely related Control Groups. In Fig. 13a, the
solitons are approaching the CGs, while Fig. 13b illustrates the
conformation of the charged states. These charges can be
neutralized by another pair of solitons of different sign with a
return to the CG configuration of Fig. 13a.

SOLITON-CHARGED   CONTROL   GROUPS

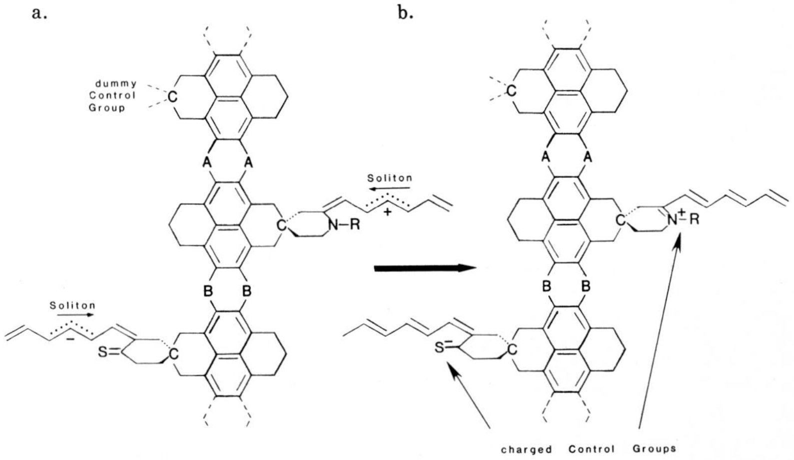

Fig. 13. A pair of oppositely charged solitons is illustrated
           approaching the Body of an electron tunnel switch in part
           a. The resulting charged Control Groups are illustrated
           in part b as linked to the Body via tetrahedral carbons
           atoms. As a further sophistication, the aromatic portions
           of the Body are linked by groups A and B which can be
           selected to control potential contours in the Body.

In regard to the design of the electron Tunnel switch Body, two additional features are illustrated in Fig. 13. First, note that the control groups are not only separated from the aromatic portion of the Body by an insulating section composed of carbon sp^3 bonds, but also that the Control Groups are twisted out of the aromatic plane by a tetrahedral carbon atom. This structure detail should help prevent the neutralization of a CG charge by the scattering and absorption of a tunnelling electron or hole. The second point of interest is the nature of the atoms A, B bridging the aromatic portions of the tunnel switch Body. Clearly, the detailed periodic potential can be adjusted in part by whether those bridging atoms are $-CH_2-$ groups, as shown, or $-CHR-$, $-O-$, $-S-$, or $-NR-$ groups.

A MECHANICAL ACTUATOR

In a recent article on "molecular" engineering, Drexler (ref. 25) reminds us of Feynman's delightful paper entitled "There's Plenty of Room at the Bottom" (ref. 26). Feynman proposes a series of miniaturizations in which small machines are used to build even smaller machines, eventually down to the molecular level. By making use of the principles of self-organization and self-synthesis as discussed herein, the chemist may be able to leapfrog the purely mechanical approach suggested by Feynman and go directly to the molecular level. However, in the spirit of a mechanical device, we suggest below a mechanical actuator controlled by the passage of a soliton.

The actuator illustrated in Fig. 14 is a combination of two ideas: The first we have seen earlier in the discussion of "smart" molecules (Fig.9), where the information state depends on the location of the labile groups L_i, which are not necessarily hydrogen (ref. 23). The second concept arises in a consideration of what happens when a soliton in trans-polyacetylene meets one or two sp^3 (or $-CH_2-$) defects (ref. 27). If a hydrogen undergoes a 1,2 shift (or 1,3 shift for two successive defects), then the soliton can pass the defect, although it may be temporarily delayed.

The mechanical device of Fig. 14a is illustrated as being actuated by a positive soliton approaching from the left. When the plus charge reaches the pyridyl nitrogen on the left, it attracts the R group, possibly (>CH-), from the opposing nitrogen and the

SOLITON MECHANICAL
ACTUATOR

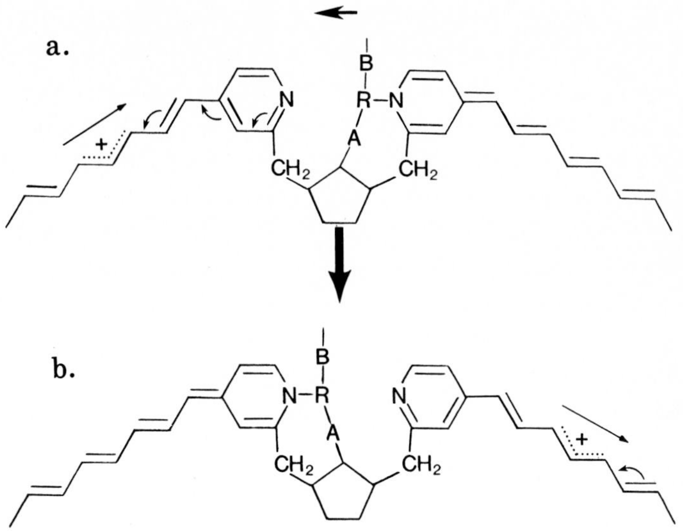

Fig. 14. This figure illustrates the use of soliton propagation to produce a motion in groups A and B of by about 2.0Å. Such a mechanical motion in the actuator group will also produce a delay in the soliton propagation.

soliton continues to the right, as in Fig. 14b. The motion of the R group by at least 2.0 Å to the left drags with it the groups A and B. Presumably, the movement or position of either A or B controls or actuates some other unspecified event at a later date. Thus we have seen how the passage of a soliton in a conjugated system can give rise to a definite group relocation by the breaking of one bond and the establishment of a second bond.

MOLECULAR MICROLASER

For current computers, it is widely appreciated that for many tasks input/output operations determine or limit the job throughput. We have stressed elsewhere that even conceptual communication with molecular switches and memories constitutes a primary barrier in MED technology (ref. 6). Of the four output modes offered (ref. 6), we review here one involving

conformational changes and solitons. This highly directional
optical output was initially suggested in 1979 (ref. 1, p. 147) in
the form of a linear array of oriented excited chromophores.
However, a technique for producing the excited state and for
triggering the photon release was not discussed. This situation
has been now remedied, as indicated in Fig. 15. In part a. to part
b. of this figure we illustrate the photoactivation of a
chromophore to an excited zwitterionic state. In the series c. to
d. to b., we see (c) excitation of the chromophores by two
approaching charged solitons, (d) rearrangement of the internal
bonding conformation of the excited chromophore by the approaching
radical soliton, and in (b) the return of the excited chromophore
to the ground state with the release of the photon. Thus the
chromophores are excited by two charged solitons and photon
release triggered by a radical soliton.

SOLITON LIGHT EMITTER

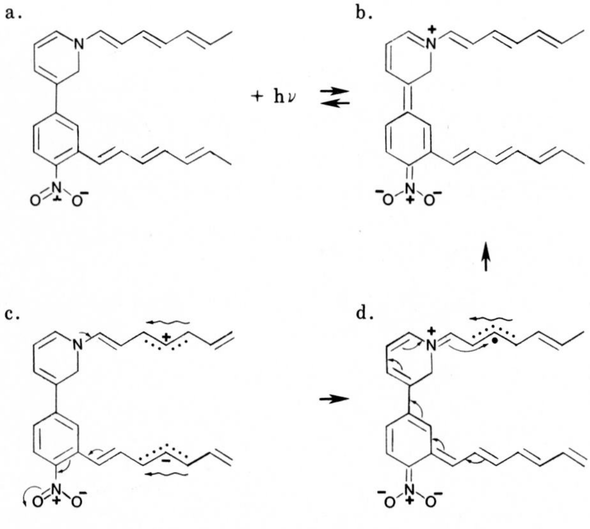

Fig. 15. Two charged solitons are employed to activate a
chromophore (c--d). The photon release is triggered by a
radical soliton (d--b--a).

Since in Section VII we indicated a proton tunnel mechanism
for the generation of pairs of charged solitons, we can now
indicate the soliton-powered molecular laser, as in Fig. 16. By

spiralling the chromophores about the vertical axis, one may be able to achieve an enhanced directionality or circular polarization. The chromophore employed in Fig. 16 is different from that of Fig. 15, however; the presumed mode of its triggered light emission (ref. 6) is related to that of Fig. 15.

SOLITON POWERED
MOLECULAR LASER.

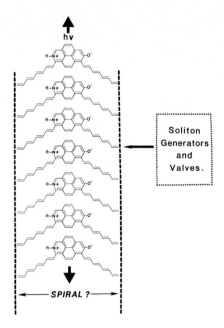

Fig. 16. An array of soliton activated chromophores might behave as a highly directional microlaser.

SOLITON AMPLIFICATION

Rightly so, engineers do not believe in dissipationless processes and instinctively require that in order to be useful, a signal must be capable of being amplified. How to achieve general amplification at the molecular level is not obvious, and the study of how amplification is achieved in biological systems should be rewarding for those interested in MED.

For electron tunnel switches using short periodic arrays, the amplification requirement is not a serious impediment since a single charge in a Control Group can turn on or off the flow of hundreds or thousands f electrons tunnelling through the Body. For solitons it could be argued that amplification of a sort is obtained in gang switches where the passage of a single soliton can turn on or off a number of different chromophores or channels (e.g., eight in the case of the completed version of Fig. 8 and Table 1).

The conceptional germ of true soliton amplification might be found in the soliton-triggered light emitter, where a photon is released via the recombination of a plus and minus soliton at a chromophore. Figure 17 illustrates the inverse process. The chromophore imbedded in the transpolyacetylene chain in Fig. 17a is photoactivated to that in Fig. 17b.

SOLITON   AMPLIFIER

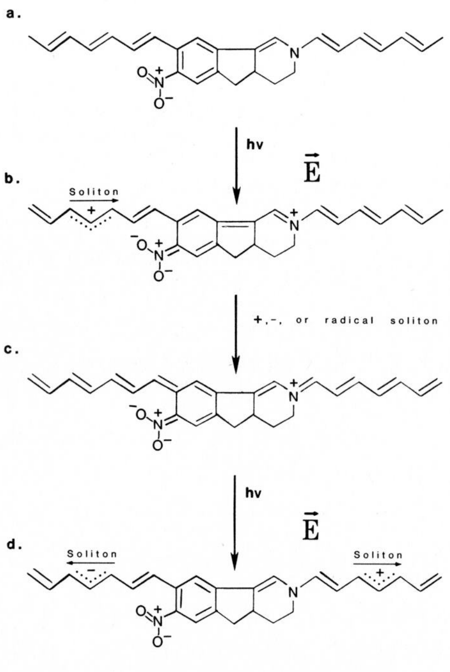

Fig. 17. This figure suggests how the passage of a soliton might be amplified by the photoactivation of a chromophore imbedded in a transpolyacetylene chain. After the chromophore has been photoactivated in part b, the passage of a soliton from left to right rearranges its internal conjugation so that it can emit two solitons as in part d.

The propagation of a radical, plus or minus soliton reorganizes a portion of the conjugated bonding in the chromophore, as in Figs. 17b and 17c. In the presence of the electric field (or possibly spontaneously), the reorganized activated chromophore in Fig. 17c emits a positive and negative soliton, propagating left and right, respectively, while the chromophore returns to the ground state (compare Figs. 17a and 17d). Thus the positive soliton in Fig. 17b, travelling from left to right, gives rise to two more solitons, shown in Fig. 17d. The energy source for this amplification is clearly the photon field, hv. Additional amplification might be achieved in the same manner by using an array of photochromes imbedded in the $(CH)_x$ chain. Figure 17 then clearly demonstrates conceptually that true solitonic amplification can take place.

MEMORY ELEMENTS

The use of molecular conformational changes can be employed to make memory elements with a gate density as high as 10^{18} gates/cc. We will discuss two such devices, one bistable (ref. 2), and the other dynamic (ref. 22). The first makes use of two different alternative valence states of a transition metal or metalloid (like tin). A small ring is formed consisting of perhaps six metal atoms with bridging ligands B (not boron) between the metal atoms. In this case, three of the metal atoms of high valence would have bridging ligands nearby (see Fig. 18a), while the ligands would be more distant from the low valent metal atoms. The high-low valence configuration is controlled by a dipole field produced by the Driver memory write device. If the Driver dipole field is reversed by soliton input, as in Fig. 18b, then the high-low metal configuration is inverted and the B ligands move toward the new high valent metal atoms. This change is propagated around the ring to the metal atom at the bottom of the ring in Fig. 18a, which metal atom serves as a Control Group for the electron tunnel switch at the bottom. Accordingly, this device provides a small bistable static memory device that can be independently written on and read.

The second memory cell is dynamic and in essence, stores charged solitons along trans-polyacetylene chains. It consists of a soliton generator, two lengths of $(CH)_x$ to and back from two-chain soliton reversers, and an electron tunnel switch whose

a.   BISTABLE CHEMICAL   MEMORY

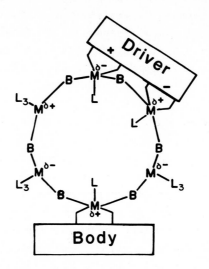

b.

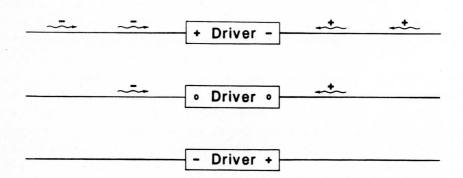

Fig. 18.   The bistable chemical memory unit of ref. 2 is shown in
          part a)   while b) illustrates the reversal of the Driver
          charges by two pairs of solitons. The   read device of the
          chemical memory is a multibarrier tunnel switch whose
          Body and one Control Group is shown at the bottom of part
          a. The   symbol B is a bridging ligand, L is another
          ligand, and L_3 is three of them.

Control Groups terminate the $(CH)_x$ return lines. The simultaneous arrival of oppositely charged solitons at the Control Groups can be used as a check against "soft" errors.

Although soliton devices may not be very fast, since the soliton propagation velocity is less than that of sound, one would be able to store information at very high densities. Assuming solitons are no closer than 200 $\overset{\circ}{A}$, and the chains are packed on a 50 $\overset{\circ}{A}$ center, the density of soliton information bits is an astounding 10^{18} bits/cc.

FUTURE DEVELOPMENTS IN MED SYNTHESIS

Synthesis is clearly a uniquely critical issue in any future that might exist for Molecular Electronic Device technology. If the necessary molecular structures cannot be synthetically realized, any feasibility argument such as this one, no matter how promising, is but a vain exercise. Therefore it is encouraging to realize that virtually any compound that can be reasonably committed to paper can be synthesized. However, it is not enough for MED purposes to have brilliant synthetic chemists available, for two primary reasons: (1) at the right time and place each reaction must in essence go to 100% completion, and (2) the ultimate product must mesh nearly perfectly with other components, not only in function but in structure. Thus, there is a space-filling and morphological requirement that must be met. These dual requirements of reaction completion and morphology are crucially coupled, but in a detailed manner of which we are only dimly aware. The self-synthesis of complex materials like cytochrome C from biological soup demonstrates clearly that the coupled requirements of reaction completion and morphology are being satisfied in the biological world. However, meeting these requirements in the synthetic laboratory is rarely attempted or achieved. Fortunately, several new chemical approaches have recently been developed that might play an important role in future MED technology.

Prior to listing opportunities for developing MED technologies in new areas of chemistry, we will give advantages that might be available in a technique not previously considered here but which might be integrated in the Merrifield-type MED synthesis, discussed earlier. The very small electrochemical cell technique is currently used as an analytical technique but might be readily adapted for MED synthesis on a substrate. The advantages of such a cell are numerous and include:

(1) Rapid exchange of solutions - that is, the amount of solution used as reagents and rinse would be very small.

(2) An electron potential can be used to drive many electric cell reactions to completion and hence avoid some of the problems suggested by the mass action principle.

(3) By reversal of current and the pole position, it may be possible to grow conducting copolymer chains of definite length depending on the number of potential reversals. Thus:

$$(\text{+metal})ABABABAB^{+} + A^{-} \rightarrow (\text{+metal})ABABABABA$$
$$(\text{-metal})ABABABABA^{-} + B^{+} \rightarrow (\text{-metal})ABABABABAB$$

(where A^{-} and B^{+} are mixed in solution but react only with the $-B^{+}$ and A^{-} end of the chain (respectively) under the right potential. This approach might be applicable to the synthesis of $-(SN)_x$ chains since many ionic species of the form $S_n N_m^{\pm Q}$ are known.

(4) Directed spatial structuring of small components should be possible using an electric field. For example, the growth of the charged polymer chain in (3) above could be achieved in an electric field so that it always stretched from left to right. In this case, the field must be reversed each time the chain end reacted and changed the sign of its charge. The use of an electric field in a small cell could be an important tool for achieving a coarse control of molecular morphology.

The central theme of this section is that new ways of thinking about chemical synthesis must be developed and practiced if MED technology is to become a reality. Areas of new chemistry that should be developed and supported in order to achieve a viable MED technology are listed below.

* The principles of self-synthesis and self-organization that apply to biological systems at the molecular level should be studied and generalized for application to non-biological chemistry.

* The principles, practice, and computational aspects of self-organization as per Stillinger and Wasserman (ref. 17) should be extended.

* Surface Modification Chemistry is central to MED synthesis.

The theories of surface reactivity and reactions of constrained groups are virtually an unknown field. Valence Bonded Surface contacts (VBS) and immunologically compatible coatings are only two of the more immediate applications.

* Very small electrochemical cells constitute a promising technology (as above).

* Laser-catalyzed synthesis offers opportunity in bond selectivity and some spatial resolution.

* Recombinant DNA techniques may become an important synthetic tool for MED components (ref. 7) if the DNA coding for non-protein products can be devised.

* Langmur-Blodgett film techniques will be useful in a variety of MED and hybrid-semiconducting technologies. However, the following chemical capabilities need to be developed: (a) a general technique for chemically bonding the LB film to the substrate and/or to other LB layers, and (b) an increased variety of techniques for bonding LB molecules together on command. Two-dimensionally ordered LB films could be useful in Molecular Lithography (ref. 6).

* Surface acoustic wave (SAW) techniques being developed at NRL (ref. 28) are capable of monitoring monolayer coverages and reactions in the presence of the vapor phase. If this sensitivity could be extended to thin liquid films and very small electrochemical cells, SAW techniques could prove to be an important tool in the development of MED technology.

The diversity of the above areas offers reasonable hope that MED synthetic problems are not insuperable; what is needed, of course, is dedication, time, synergism, and, not least, support.

CLOSURE

In our discussion we have attempted to indicate both the great variety of conformation switching phenomena available, as well as synthetic challenges inherent in the future of Molecular Electronic Device technology. It is clear that if these challenges, and the challenges to the semiconducting industry, are to be met, we must bring some form of closure to the concept and promise of molecular engineering. This latter concept has been with us since the early sixties, but has not progressed beyond the traditional chemist's batch process. Certainly new approaches or combination of approaches, such as we have advocated above, are required - and surely will come - to make true MED technology a reality.

The MED concepts, as indicated above, are truly exciting and surely worthy of pursuit for their own sake; however, they potentially impact not only the semiconducting industry and most every area of chemistry but many aspects of human activity, such as communication, industry, health, defense, surveillance, data storage, to name just a few. While it is unlikely that the molecular switching devices will come to pass in the form described above, it is certain that "molecular" chemistry will be vital to the development of future computers. What better place to start the vitalization of chemistry than among the educators and conferees of this congress.

REFERENCES

1 F. L. Carter, "Problems and Prospects of Future Electroactive Polymers and 'Molecular' Electronic Devices," in the NRL Program on Electroactive Polymers, First Annual Report, Ed. L. B. Lockhart, Jr., NRL Memorandum Report 3960, p. 121.

2 F. L. Carter, "Further Considerations on 'Molecular' Electronic Devices," in the NRL Program on Electro-active Polymers, Second Annual Report, Ed. R.B. Fox, NRL Memorandum Report 4335, p. 35.

3. F. L. Carter, Ed., "Proceedings of the Molecular Electronic Device Workshop," held 23-25 March 1981, NRL, Washington, D.C., NRL Memorandum Report 4662, October 1981; also to be published by Marcel Dekker, New York, N.Y., Fall, 1982.

4   F. Aviram and M. A. Ratner, Chem. Phys. Letters, 29, (1974) 277.

5   F. L. Carter, "From Electroactive Polymers to the Molecular Electronic Device Computer," in "VLSI Technologies -- Through the '80's and Beyond," Eds. Denis McGreivy and Kenneth Pickard, IEEE Computer Society, No. 424, Washington, D.C., (1982), p. 328; and in "Proceedings of Winter School on Electroactive Polymers," held 3-10 January 1982, Font Romeu, France, Ed. P. Bernier, CNRS (in press).

6 F. L. Carter, "Toward Computing at the Molecular Level," NATO Advanced Research Institute at Les Deux Alpes, France, 22-26 March 1982, in "Microcircuitry-Structure and Complexity," ed. R. Dingle, in press.

7 E. M. Ulmer, "Biological Assembly of Molecular Ultracircuits,"
   In Ref. 3, p. 167.

8 C. Birr, Aspects of the Merrifield Peptide Synthesis,
   Springer-Verlag, Berlin, Heidelber, New York, 1978.

9 A. G. MacDiarmid and A. J. Heeger, "Recent Advances in the
   Chemistry and Physics of Polyacetylene; Solitons as a Means of
   Stabilizing Carbonium and Carbanions in Doped $(CH)_x$," Ref. 3,
   p. 208.

10 O. J. Scherer and R. Wies, Z. Naturforsch., 25b (1970) 1486.

11 J. W. Milliken, "Metallic Covalent Polymers: Derivatives of
   $(SN)_x$ and $(CH)_x$," Chemistry, Ph.D. Dissertation, Univ. of Pa.,
   1980.

12 R. W. Murray, Acc. Chem. Res. 13 (1980) 135.

13 E. A. Pshenichnov, Soviet Physics - Solid State, 4 (1962) 819.

14 F. L. Carter, "Electron Tunnelling in Short Periodic Arrays,"
   Ref. 3, p. 344; also NRL Memorandum Report 4717 (1981).

15 A. Aviram, private communication.

16 P. M. Kuznesof, K. J. Wynne, R. S. Nohr, and M. E. Kenney,
   J.C.S. Chem. Comm., 121 (1980); R. S. Nohr, P. M. Kuznesof, K.
   J. Wynne, and M. E. Kenney, J. Am. Chem. Soc. 103 (1981) 4371.

17 F. H. Stillinger and Z. Wasserman, J. Phys. Chem. 82 (1978)
   929.

18 A. S. Davydov andN. I. Kislukha, Phys. Stat. Sol. (b) 59 (1973)
   465; Sov. Phys. JETP, 44 (1976) 571.

19 W. P. Su, J. R. Schrieffer, and A. J. Heeger, Phys. Rev. Lett.
   42 (1979) 1698; Phys. Rev. B22 (1980) 2099.

20 S. Kivelson, Phys. Rev. Lett. 46 (1981) 1344.

21 D.K. Campbell and A. B. Bishop, Nuclear Physics B200 (FS4)
   (1982) 297.

22 F. L. Carter, "Conformational Switching at the Molecular
   Level," in Ref. 3, p. 53.

23 R. C. Haddon and E. H. Stillinger, "Molecular Memory and
   Hydrogen Bonding,: in Ref. 3, p. 17.

24 A. van Dormael, Chim. Ind. 67 (1952) 368; Compt. Rend. 23
   Congr. Intern. Chim. Ind., Milan (1950).

25 K. E. Drexler, Proc. Natl. Acad. Sci., USA, 78 (1981) 5275.

26 R. Feynman, "There's Plenty of Room at the Bottom," in
   Miniaturization, Ed. H. D. Gilbert, Reinhold, New York (1961),
   p. 282.

27 F. L. Carter, "Solitons and SP3 Defects in Trans-
    Polyacetylene," in "Proceedings of Winter School on
    Electroactive Polymers," ed. P. Bernier, CNRS, France, in
    press.

28 H. Wolhtzen, private communication.

S.R. Heller and R. Potenzone, Jr. (Editors), *Computer Applications in Chemistry*
© 1983 Elsevier Science Publishers B.V., Amsterdam — Printed in The Netherlands

COMPARISON OF LOGICAL OR THREE-DIMENSIONAL MOLECULAR STRUCTURES BY MEANS OF AN AUTO-CORRELATION PROCEDURE

PIERRE BROTO, GILLES MOREAU AND CORINNE VANDYCKE
Centre de Recherches ROUSSEL UCLAF, BP 9, 93230, Romainville, FRANCE

INTRODUCTION

In this talk I am going to present our methods and the state of advancement of our work on the comparison of molecules, a fundamental problem in drug design. If medicinal chemists are interested in such comparisons, it is obviously because similar molecules might have similar behaviour. Not only the chemical behaviour is meant, but also and perhaps more often the non-bonded interactions which are responsible for phenomena such as transport, recognition and affinity.

In the first part we define what we are going to compare, namely our perception of a molecule as a distribution of properties on a structure, either Logical or 3-dimensional. In the second part the question of the descriptor is treated : what we want it to be is intrinsic, in other words dependent only on the molecular structure and nothing else. This is obtained by a kind of autocorrelation procedure. In the third part some examples are given where these descriptors are used in Structure Activity Studies. (refs. 1, 2)

PERCEPTION OF A MOLECULE

Traditionally there are two ways of perceiving a molecule : the most common is probably the so-called **Structural Formula,** the other is the **molecular shape,** that is to say the volume occupied by the molecule in space. These two conceptions are frequently opposed, the second being presented as more realistic than the first. In fact we will see that the difference is not so great.

The Structural Formula

On the structural formula we recognize some _codes_ such as the bond lines, the atomic symbols, rings and chemical functions, etc... In fact, when a medicinal chemist compares two structural formulae, he implicitly makes use not only of structural information, but also of physical properties associated with the symbols and substructures such as chemical functions. This perception which is still fuzzy for many people can be rationalized as follows:

The Logical Structure

Figure 1 represents the structural formula of chloramphenicol. Firstly, it is a graph : the nodes are the atoms and the links are the bonds. Both are independent of their chemical and geometrical properties. The aim of this graph is just to tell us which atoms are connected together.

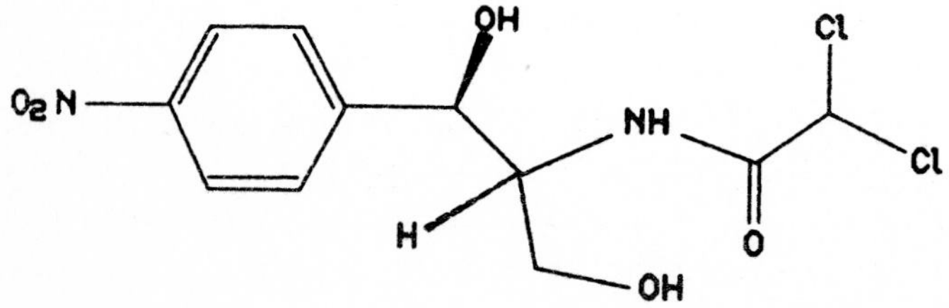

Fig. 1. Structural formula of chloramphenicol.

Secondly, to complete the definition of this molecule we only have to **distribute** on these nodes the following properties = atomic number and functionality, which allows us to find out where the double bonds are located. Even the chirality can be expressed by a distribution. Furthermore if we consider that sp2 atoms can be asymmetrical (in their nodal plane) (ref. 3) then all the structural information contained on a structural formula can be thought of as distributions of structural properties on the nodes of the graph.

It is interesting to note that the graph need not respect any particular shape provided the relations of neighbourhood are correctly expressed. This means that the structural formula is not merely a two-dimensional structure, as some people wrongly claim.

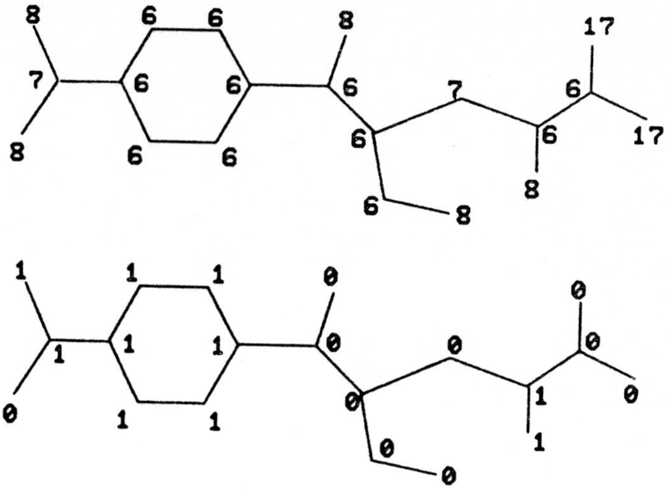

Fig. 2. Distributions of "Atomic Number" (above) and of "PI Functionality" (below).

The Logical Space

The ultimate step in the rationalization of the structural formula is to set up the graph as a **Logical Space** : this space consists of a finite number of points, the nodes. The distance between two nodes is defined as the smallest number of links between them. Such a definition of distance satisfies the 3 usual requirements:

$$d(i,i) = O$$
$$d(i,j) = d(j,i)$$
$$d(i,j) < d(i,k) + d(k,j)$$

Now the structural formula appears as a mathematical entity, namely the distribution of properties in logical space.

Spatial Structure and Molecular Shape

The spatial structure consists of the atoms, (points) located in 3-D space by their coordinates x, y and z. As in the case of the logical structure we can associate each atom with a property such as connectivity, (n) functionality etc...

For obvious reasons the most suggestive distribution is that of the atomic volumes : its graphical representation is the traditional molecular shape (see Figure 3).

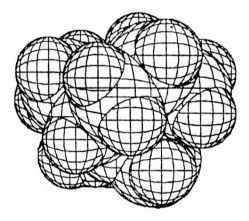

Figure 3.

Now the analogy between the Logical Structure and the 3-D Structure is obvious, only the nature of the representation space is different.

Generalization

Distribution of Physical Properties

A molecule is not a uniform and homogeneous object ; its properties change from point to point. We suggest that we study the distribution of several properties, either structural or physical, on its structure, either Logical or 3-dimensional.

The following list contains the most common properties we currently use for Structure Activity Studies.

- Connectivity (measured by 1, 2, 3, 4 ...)
- Electronegativity (Pauling's Scale)
- Van der Waals Volume
- π Functionality (0, 1 or 2)

- Hydrogen bond donor ability (0 or 1)
- Hydrogen bond acceptor ability (0 or 1)
- Atomic Contribution to LogP octanol water

This last property deserves some explanation.

The partition phenomena are perhaps the most direct manifestation of non-bonded interactions. Our idea was that it would be useful to know how the interaction potentiality is spread over the structures. Thus we designed a system of ("additive and constitutive") atomic contributions to LogP. In Figure 4 are shown the pieces of information taken into account in the contributions.

Carbon atoms, which are considered to be not very polarisable, are characterized by their bond environment only. Heteroatoms, which have lone pairs of electrons, are given a more extended environment that takes into account conjugation with double bonds and with alphaheteroatoms (Whatever their nature is).

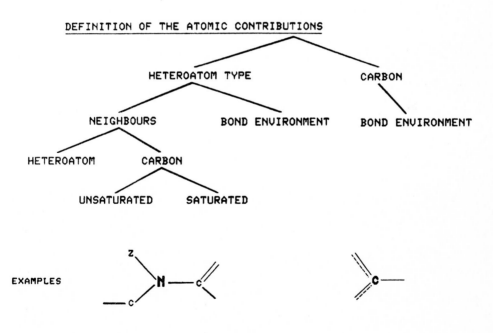

Figure 4.

The design of this system of contributions was also dependent on (and constrained by) the data available : a set of 1500 experimental LogP in the octanol-water system, for molecules without internal hydrogen bond. These data were compiled by Hansch and coll.(ref. 4) Each molecule was described by a linear equation. The system of 1500 equations and about 200 unknowns was solved by means of a Monte Carlo procedure. Figure 5   shows some structural contributions to LogP. The two numbers in the upper part of each box are respectively the number of molecules in the data set that contains the substructure and the number of times it is found in the equation system. The greater these numbers are, the more reliable the contribution is.

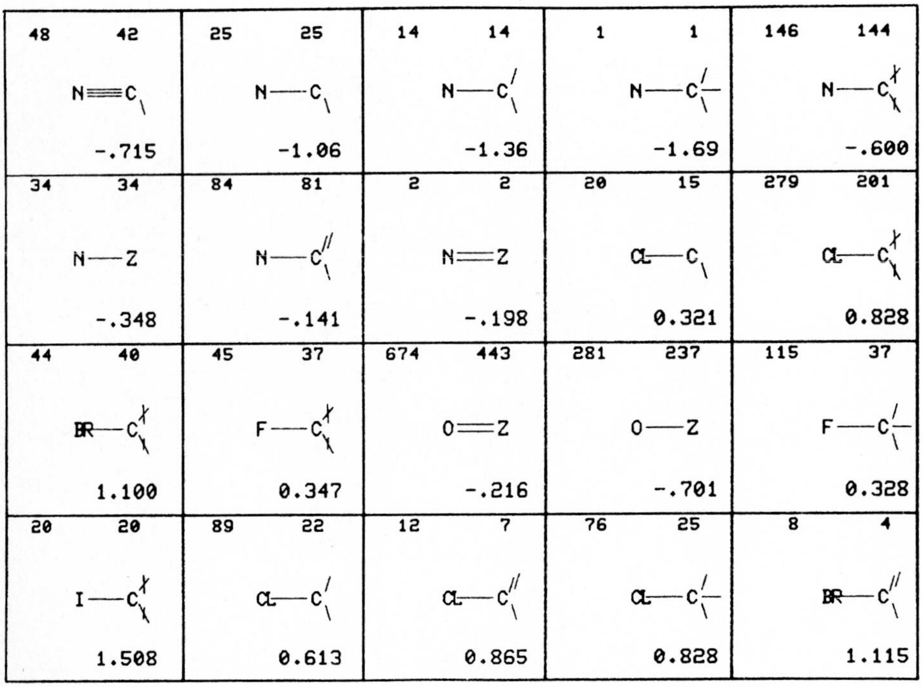

48 42	25 25	14 14	1 1	146 144
N≡≡C -.715	N——C -1.06	N——C -1.36	N——C -1.69	N——C -.600
34 34	84 81	2 2	20 15	279 201
N——Z -.348	N——C″ -.141	N≡≡Z -.198	Cl——C 0.321	Cl——C 0.828
44 40	45 37	674 443	281 237	115 37
Br——C 1.100	F——C 0.347	O≡≡Z -.216	O——Z -.701	F——C 0.328
20 20	89 22	12 7	76 25	8 4
I——C 1.508	Cl——C 0.613	Cl——C″ 0.865	Cl——C 0.828	Br——C″ 1.115

Figure 5.

MOLECULAR DESCRIPTOR

Let us examine how molecular shapes are classically compared : usually only molecules belonging to the same series are compared ; for instance steroids, so that it is more or less postulated that ring A (on the first molecule) should be superimposed with the same ring on the second molecule. A measure of dissimilarity is given by the sum of the squares of the distances between homologous atoms. Such a procedure is **arbitrary** and **limited** to objects of the same family. It cannot be extended to comparing very different molecules.

An analogous difficulty is encountered with the logical structures comparisons. The Free Wilson and Darc Pelco methods are limited to molecules having the same logical substructure : the common substructure acts as a common referential.

Therefore the question is : is it possible to describe a structure either logical or spatial, independently of any referential.

The answer again comes from mathematics. Figure 6 shows what is called an autocorrelation function ; $F(t)$;

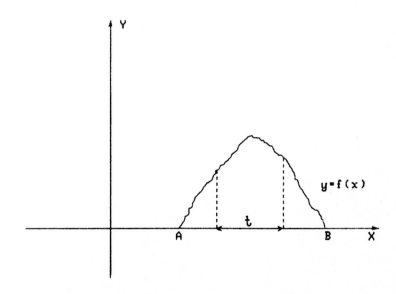

AUTOCORRELATION FUNCTION $F(t) = \int_{AB} f(x).f(x+t)dx$

Figure 6.

For instance x may be time and f(x) a signal depending on time.

The main feature of F(t) is that it does not depend upon the location of the AB interval along the x-axis (it is obvious).
The autocorrelation function F(t) is a description of the original signal, which is a distribution of numerical values on the continuous space A-B. F(t) is less accurate than f(x) and generally it is not possible to go from F(t) back to f(x).

The process of autocorrelation associates in the same summation all the values taken at the same separation t. It may be itself. F(t) is independent of the referential.

Now we have just to transpose this procedure in the cases of the Logical and 3-D structures.

<u>Autocorrelation of a distribution on the Logical Structure</u>

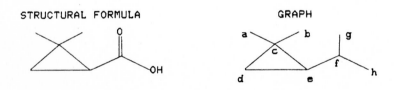

STRUCTURAL FORMULA   GRAPH

FROM	PATH LENGTH				
	0	1	2	3	4
a	f(a).f(a)	f(a).f(c)	f(a).f(b) f(a).f(e) f(a).f(d)	f(a).f(f)	f(a).f(g) f(a).f(h)
b	......	......	......	......	.....
c	f(c).f(c)	f(c).f(a) f(c).f(b) f(c).f(e) f(c).f(d)	f(c).f(f)	f(c).f(g) f(c).f(h)	
:	......	....			
	$\sum_0$	$\sum_1$	$\sum_2$	$\sum_3$	$\sum_4$

Figure 7.

On Figure 7 we find in one column all the pairs of atoms separated by a specific distance 0, 1, 2 etc...For instance atom $\underline{a}$ has only one first order neighbour, but 3 second order neighbours. These atom pairs are the analogues of the pairs of points x and x+t in Figure 6, except that in Figure 6 the distance t varies continuously whereas the Logical distance varies discretely.

Let f(i) be the measure of an atomic property attached to node $\underline{i}$; if we calculate all the products (as f(a).f(b) for the pair $\underline{a}$ $\underline{b}$) and if we add up these terms along a column we get a series of terms symbolized by E_0 to E_4. This summation is the analogue of the integration at constant t, in Figure 6.

The Es are the components of a vector, the autocorrelation of f defined on the logical space.

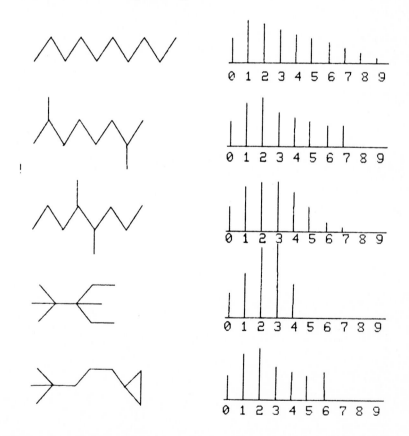

Figure 8.   Autocorrelation Vector Components, with F(I) = Van Der
            Waals Volume.

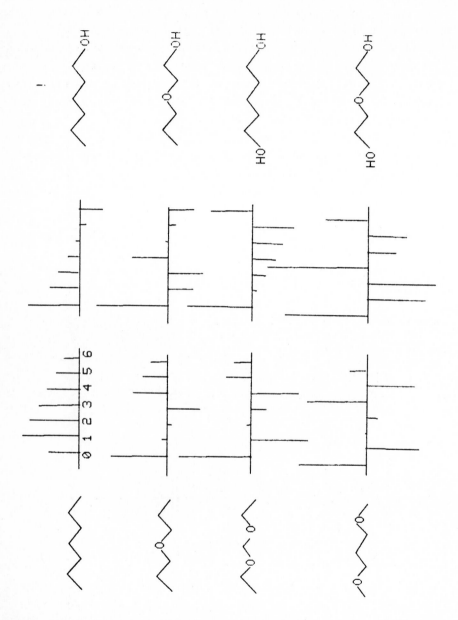

Figure 9. Autocorrrelation Vector Components, with F(I)=Atomic
          Contribution to LOGP.

The autocorrelation function is an intrinsic property of a molecule. There is in this procedure a loss of information, but it is perhaps the price we have to pay for independency. With such descriptors very different molecules can be compared : components of the same rank correspond to products associated with the same pathlength, wherever they are in the molecules.

Figure 8 shows some hydrocarbons that have the same number of carbon atoms but different graphs. The autocorrelation vectors for the property Van der Waals volume are represented as histogrammes (each peak is a component of the vector). As all the atoms are carbons, the f(i) values can be considered to be a constant, so that the components of the autocorrelation vector are proportional to the number of paths of length 0, 1, 2 ....

All the molecules of Figure 9 have the same skeleton, but the number and location of oxygen atoms vary. The property used in this case is the contribution to octanol-water LogP (about 0,5 for carbons and -1,.. for oxygens). The autocorrelation vector are much more complex than in the preceding example.

Everybody will easily find that with this property the sum of the components is the square of LogP.

Autocorrelation of a distribution on a 3-D Structure

The transposition of the procedure to consider 3-dimensional structures is almost obvious. In order to change we describe it under the form of an algorithm. Let the atomic property under consideration be f. The horizontal axis on Figure 10 is a scale of distance divided in small intervals, say for instance 0.2 Å. Each interval is given an order number, for example distances between 5 and 5.2 Å fall in the 26th. interval.

- select a pair of atom   i.j
- calculate the distance   d(i,j)
- calculate the product   f(i).f(j)
- add this product to the accumulator correcponding to the interval in which d(i,j) falls.
- repeat the process for all the pairs of atoms.

The spatial autocorrelation function is thus obtained ; Figure 10 shows it in the form of an histogramme. Obviously the function is 0 for distances between 0 and about 1.3 Å because there is no such interatomic distance.

Once again these descriptors allow the comparison of molecules

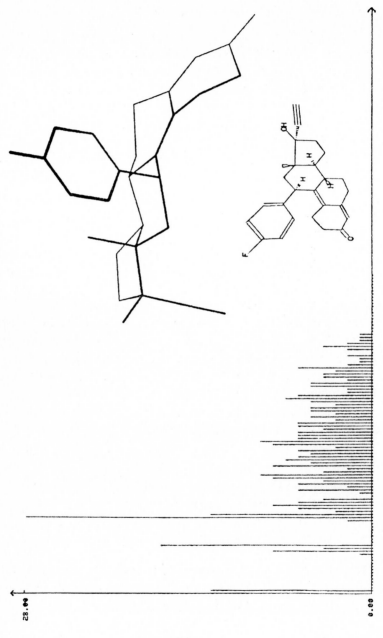

Figure 10.

without having to decide how they have to be superimposed. Any physical or chemical property within our reach can be used.

In the computer these autocorrelation functions are again vectors, each component corresponding to a distance interval. Usually we use $0,2$ Å to $0,5$ Å intervals. Leading to vectors of about 20 to 50 components for molecules of the size of steroids.

APPLICATIONS, EXAMPLES

Some observations on the use of these descriptors

Firstly when we have a set of molecules there is at least one which is the "smallest" (the size of a molecule is the largest interatomic distance either logical or classical). All the vectors are truncated at the largest common component order.

Secondly a descriptor can be composed of several truncated autocorrelation vectors as represented in Figure 11.

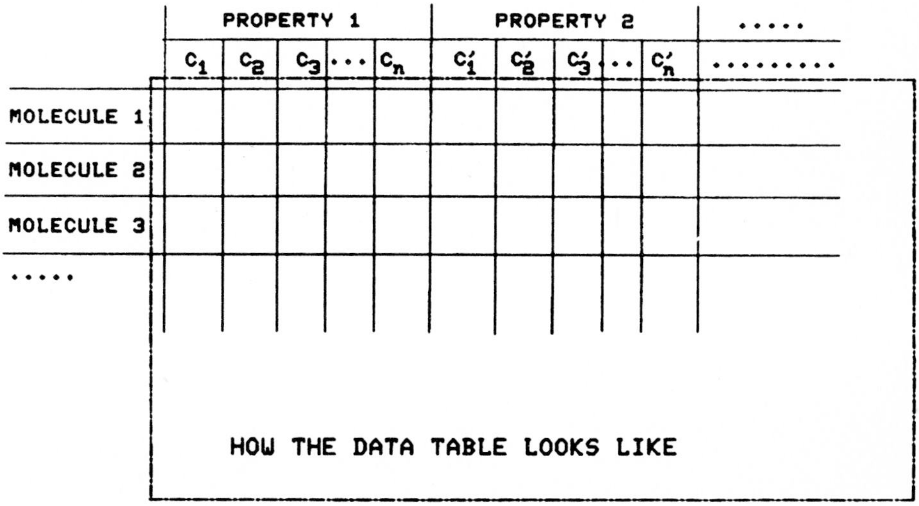

	PROPERTY 1				PROPERTY 2				. . . . .
	c_1	c_2	c_3	$\cdots c_n$	c_1'	c_2'	c_3'	$\cdots c_n'$	. . . .
MOLECULE 1									
MOLECULE 2									
MOLECULE 3									
. . . . .									

HOW THE DATA TABLE LOOKS LIKE

Figure 11.

Thirdly the most appropriate methods for handling such an amount of data are the so-called "data analysis techniques" such as the multivariate methods and factorial analyses.

Fourthly, after having extensively used methods like KNN (K nearest neighbours) and LM (learning machine) we now proceed first with a PCA (principal component analysis) (ref. 5) just to examine

the structure of data and then we use, for instance, a
discriminant analysis programme. This examination with PCA is
necessary because of the high correlations between the vector
components.

Glafenines and Indomethacines. (ref. 2)

Glafenines (Figure 12) are peripherally acting analgesics. A
set of about 150 such molecules described by their logical
structures were autocorrelated using the classical properties
already mentioned. This leads to a data table as in Figure 11.
This table is submitted to a PCA in order to get a map of the set
of points (molecules).

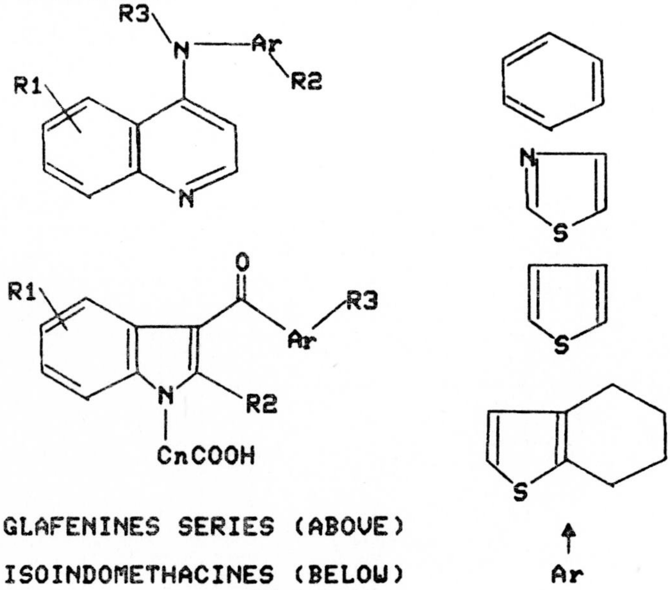

GLAFENINES SERIES (ABOVE)

ISOINDOMETHACINES (BELOW)   Ar

Figure 12.

The first factor (horizontal axis) accounts for about 50% of
the total variance, and the true dimensionality of the cloud does
not exceed 10, because 90% of the variance is accounted for by the
first 10 factors.

The map of Figure 13 is the projection of the cloud on the F1
F2 plane. Active molecules are in red and the inactive ones are in
blue (also represented by the letters A and I). Fortunately, and
also perhaps fortuitously, this plane shows a fairly good
dichotomy between the two activity classes. It allows us also to
see that the set of molecules is almost homogeneous and therefore

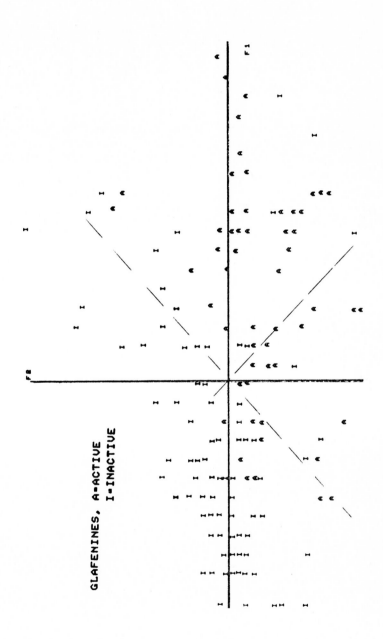

Figure 13.

all the set can be handled at a time.

If we perform a discriminant analysis (DA) on this set we get a recognition percentage of about 89% and a good prediction percentage of about 86%, with only 3 variables in the discriminant function.

However let us consider Figure 14. It represents the projection of the initial variables on the same F1-F2 plane. We see that all the electro-negativity variables project in the same region (they are correlated) far from the origin and almost on the dichotomic direction. We can therefore expect this property to be found in a discriminant analysis and actually it is selected first. In contrast the π functionality variables (which are also highly correlated) project very far from the direction of dichotomy, indicating that this property has no discriminating

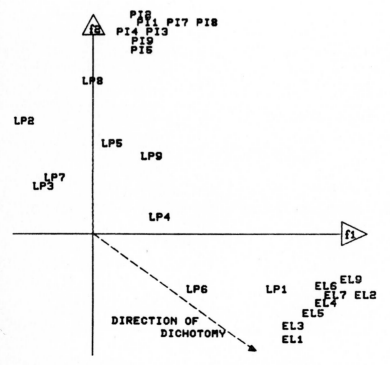

Figure 14. Projection of the Variables on the f1-f2 plane.

power, fig. 15 shows the results in recognition and prediction if
only π functionality is used : the results obtained are very
poor since the percentages remain at above 50%, that is to say no
prediction at all.

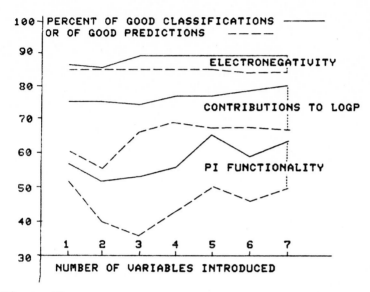

Figure 15.

The variables associated with the atomic contributions to LogP
project almost everywhere but in the center of the map ; these
variables are not highly correlated, and we can expect them not to
be very discriminant, and   this is actually shown by the
discriminant analysis with this property, Figure 15.

The same map is shown on Figure 16, a set of indomethacine
derivatives were added and positioned with respect to the factors
Fl and F2. This is a way of seeing whether or not this series can
be mixed with the glafenines. This mixing seems reasonable and
this new series can actually be predicted with a fairly good
percentage of success (about 73%) when only the glafenines are
used in the training set of the DA.

Of course, much better results (80-83%) are obtained when both
series are used in the training set.

Such a technique is now currently used in our laboratory to
investigate new leads in the field of agrochemicals.

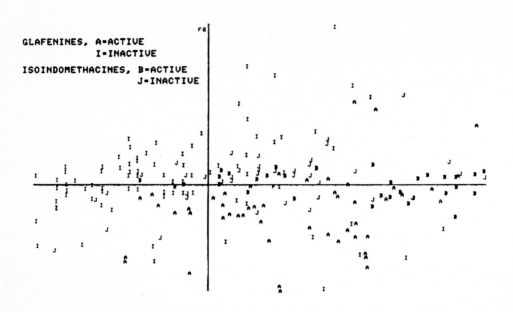

Figure 16.

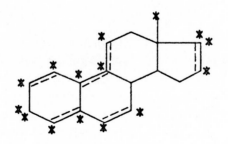

✻ : SITES OF SUBSTITUTION

‗‗‗ : DOUBLE OR SINGLE BONDS

Figure 17.

Steroids

In this study, which is still in progress, the 3-D structure is used. About 175 steroids tentatively represented on Figure 17 were calculated in three dimensions, in their lowest energy conformation by means of a molecular mechanic programme. These molecules were tested for their affinity for the glucocorticoid receptors of rat thymus, and divided into two classes = high and low affinity classes (threshold at about 20% of the dexamethasone affinity).

For the calculation of the autocorrelation functions (with VdW volume and contribution to LogP) we chose distance intervals of 0,2, Å and also 0,33 Å.

As in the first example our initial task is to operate a PCA (principal component analysis) to visualize the "cloud" of points. Whatever the property is, it appears that the real dimensionality of the cloud is much higher than with the logical structure. For instance the first two principal components account for only 20% of the variance.

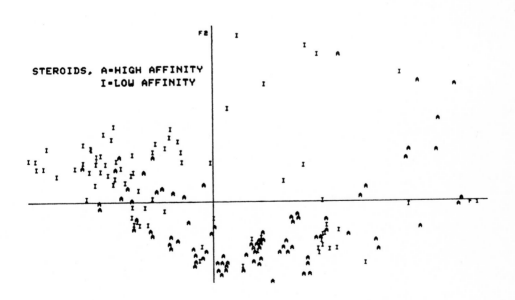

Figure 18.

Figure 18 shows the Fl-F2 plane for the property VdW volume (namely the shape of the molecules). On this map high affinity molecules are in red and low affinity ones in blue. No dichotomy is obvious. A discriminant analysis of this cloud leads to only 75% correct recognition.

Now if we autocorrelate the distribution of the LogP contributions we get the surprising picture (Fl F2 plane) of Figure 19.

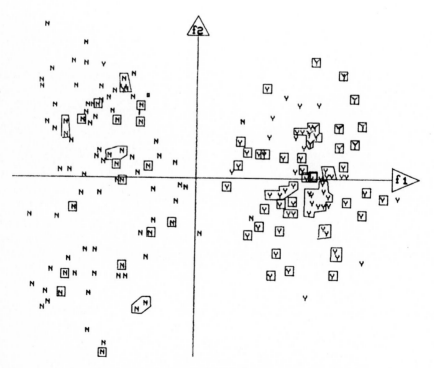

'Y' : STEROIDS WITH 3-KETO,Δ4,Δ9-10 SUBSTRUCTURE.

▱ HIGH AFFINITY CLASS

Figure 19.

The cloud is split into two parts ; a fairly good partition (high affinity/no affinity) appears ; it corresponds to 77% of correct classification (once again it is not the purpose of PCA to show good dichotomies). If we apply a DA on this data we get an 81% correct classification. This DA can either be applied on the initial variables or on the Principal Factors ; in the second case factor F1 plays an important role in the discriminant function.

A rapid examination of the individual molecules shows that the two subclouds of Figure 19 correspond to the two classes of molecules = those containing the 3 keto Δ4 Δ9-10 substructure and those which do not have this substructure. Furthermore, even if it is slightly disappointing, we are almost convinced that the 81% correct classification according to the affinity class, is in fact explained (contained) by the almost perfect recognition of the 3 keto Δ4 Δ9-10 substructure. (That is to say, if we state as follows : a molecule that contains 3 keto Δ4 Δ9-10 has a high affinity, otherwise it has low affinity, then we get about 77% correct classification).

We repeated these analyses with a different distance interval, namely 0,33 $\overset{\circ}{A}$. We find the same dichotomy as before furthermore, we also see a third subcloud on the left side and at the bottom of the diagram. (Already visible on Figure 19).

After examination of the structures associated with this cloud, we found that it corresponds to steroids having an A/B cis junction either because of a 5α-10α epoxide, a 10α-9α epoxide, or a structure with the 10β methyl group but with a 5β substituent or hydrogen.

Experiments with these structures are still in progress. If for the time being we have not really succeeded in correlating the affinity with the structure, I think that we have at least a good tool to recognize shapes and distributions and to classify them.

CONCLUSION

In this article we have presented a new perception of a molecule and a new molecular descriptor. They present remarkable features such as :
- either the logical structure or the 3-D structure can be used, in the same way.
- intrinsicalness.
- finally, they allow us to compare molecules from the point

of view of the distribution of one or several properties at a time.

They have already proved to be useful in our hands ; we hope they could be of great help in finding new leads in medicinal chemistry.

Some experiments indicate they might be useful too in questions such as the study of relationship between spectra and structures.

REFERENCES

1 G. Moreau and P. Broto, Nouveau Journal de Chimie, 4 (1980) 359.

2   G. Moreau and P. Broto, Nouveau Journal de Chimie, 4 (1980) 757.

3 E. J. Corey, W.J. Howe and D. A. Pensak, J. Amer Chem. Soc., 96 (1974) 7724.

4   C. Hansch and A. Leo, Pomona College Medicinal Chemistry Project, Claremont, California.

5 We use the data analysis programmes of ADDAD (Association pour le developpement de l'analyse des donnees), 4 place Jussieu, 75005 PARIS, France.

S.R. Heller and R. Potenzone, Jr. (Editors), *Computer Applications in Chemistry*
© 1983 Elsevier Science Publishers B.V., Amsterdam — Printed in The Netherlands

COMPUTER-ASSISTED DRUG DESIGN

Y.C. MARTIN, K.H. KIM, T. KOSCHMANN, T. J.O'DONNELL
Pharmaceutical Products Division, Abbott Laboratories, North Chicago, IL, 60064

INTRODUCTION

In this chapter we will describe how we use computers to help chemists decide which compounds to synthesize as potential new drugs. We will concentrate on the kind of problem typically faced by medicinal chemists at Abbott-- that is, that a "lead" compound with interesting biological activity has been identified but there is no structural information on the target biomolecule with which it interacts. This means that we must plan our theoretical and synthetic strategy using structure-activity information on the small molecule only.

As an example, several years ago we discovered that the 5'-carboxamide of adenosine, I, increases the blood flow to the

heart dramatically and for a prolonged period compared to adenosine (ref. 22). Accordingly, a number of other 5'-substituted analogues were prepared and evaluated in the same dog test. Of what use is the computer in designing these analogues and in analyzing the resulting structure-activity information?

The strategy is to prepare and test a set of analogues of the lead compound so as to explore in a systematic fashion the requirements for biological activity. The exact set of molecules to be synthesized is planned with the aid of computer methods, as is the resulting structure-activity information.

Two important decisions are made by the theoretician. The first is how to transform a <u>molecular</u> <u>structure</u> into a <u>numerical</u> <u>description</u> of a molecule. Of course, it is important that this numerical description contain all of the information that characterizes its tendency to interact with the target biomolecule(s). In particular, the molecules may be described by several quantitative descriptors that reflect their ability to participate in noncovalent interactions. Once one has measured the biological response and calculated the structural properties of the molecules, the second critical decision is how to evaluate the relationship between these descriptors and the target biological response. Once a relationship is known, it is used to predict the response of as yet untested compounds.

We will briefly describe the application of two types of strategy to the adenosine analogues. The first strategy is QSAR. It describes the molecule by the properties of the variable substituent only, and so its description of the three-dimensional conformation and shape is indirect. The advantages of QSAR are that it is fast and easy to perform and that the analysis of the relationships between compounds is a straightforward statistical problem. On the other hand, one may choose to describe molecules more accurately by consideration of the three-dimensional position and movement of atoms and electrons in space. The trade-off for this more accurate molecular description is that the mathematical techniques to analyze the structure-activity relationships between molecules are not available and one must usually rely on human pattern recognition with its inherent possibilities of bias. Each computer-assisted drug design problem requires the appropriate balance between accurate molecular descriptions and objective statistics. To us this means that more than one strategy should support any conclusion, and that theoretical conclusions should be considered to be hypotheses to be used to suggest new molecules for synthesis.

CALCULATION OF SUBSTITUENT EFFECTS OF INDIVIDUAL MOLECULES

The simplest yet very powerful application of computers in drug design is represented by QSAR (refs. 10, 15). Values of many of the substituent constants have been collected by (ref. 11) and are also available in computerized form.

Hydrophobic interactions result from the tendency of nonpolar molecules to associate with each other rather than water. The driving force comes from the strong intermolecular interactions of water. Because the structure of water is so poorly understood in quantitative terms, one cannot reliably calculate hydrophobic properties from first principles. However, a reliable and convenient descriptor of hydrophobicity is the logarithm of the octanol-water partition coefficient.

Approximately twenty years ago Iwasa, Fujita, and Hansch demonstrated that such a logP value is an additive property, and coined the term π value for such hydrophobic substituent constants (ref. 7). Although in principle it is possible to calculate logP from molecular structure, we generally consider a set of related molecules and so we use π values (or the newer f constants) in our calculations.

The Pomona group maintains a computerized data base of all measured partition coefficients. The structures are described by Wiswesser Line Notation code as well as common name and molecular formula. While we usually search a print-out manually, we sometimes use a text editor on the PDP-10 to do so.

Table 1 lists the structure of the R2 substituent and the calculated physical properties of the set of 5'-carboxamides of adenosine to be considered.

Dispersion or induced dipole-induced dipole interactions are generally quantified by the molar refractivity of the variable substituent.

Electrostatic interactions include charge-charge, charge-dipole, and dipole-dipole interactions. In the simplest cases one can use a Hammett σ constant to describe the effect of the variable substituent on the electronic properties of the (remote) constant portion of the molecule.

However, sometimes it is necessary to consider descriptors derived from quantum chemical calculations. For example, Boyd et al., used a measure of transition state energy derived from CNDO/2 calculations in a QSAR of cephalosporin analogues (ref. 2).

Steric repulsion in the simplest cases can be described by the dimensions of the variable substituent only. The most familiar measure of substituent size is the Taft E_s value. For such cases one can use the simple tabulated values for the length (L), and minimum and maximum diameter of the substituent, B_1 and B_4, respectively (ref. 24).

Conformational properties other than steric hindrance can of course also influence the relative potency of an analogue. An added group can change the preferred conformation or the conformational flexibility of a molecule. In such cases one can use either quantum chemical or potential energy calculations to investigate the possibilities for each analogue. The application of such calculations to the adenosine series will be discussed below.

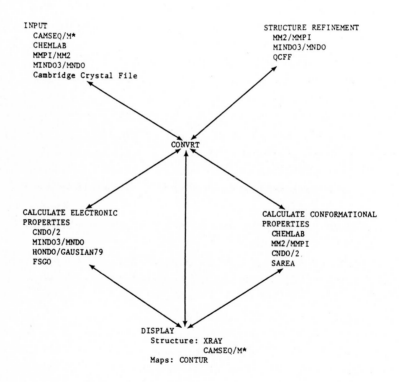

* on Tektronix 4054

Fig. 1. Relationship of computer programs in the Abbott Interactive Molecular Manipulation and Display system.

CALCULATION AND DISPLAY OF THREE-DIMENSIONAL PROPERTIES OF MOLECULES

The AIMMD (Abbott Interactive Molecular Manipulation and Display) system, Fig. 1, is a loosely joined collection of programs for molecular modeling. The programs were for the most part written elsewhere and were acquired through facilities such as the Quantum Chemistry Program Exchange (Indiana University). Most programs are currently run on a DECsystem-10. A TEKTRONIX 4054 microprocessor is used to run the molecular modelling program CAMSEQ/M. Some calculations are performed on the IBM mainframe.

CONVRT was written at Abbott because there are no conventions for the format of input files used and output files generated by these externally developed programs. For example, it is usually inconvenient to use the results of a geometry refinement calculation as input to a graphic display routine. CONVRT is a file format interconversion program. It uses output files from any program in AIMMD to prepare the input files for other programs in the system. It politely prompts the user for information that is not supplied in the input file and which it cannot calculate. It also does such things as add lone pairs for input into MM2 calculations and strip off these lone pairs when MM2 output geometry is used as input into potential energy or quantum chemical calculations. The advantage of using the CONVRT strategy is that we do not change the program written by the experts, and when they supply an updated version we merely install it. The second advantage is that we do not have to define one type of data file that must contain every type of information that any future program might need to know about a molecule. The disadvantage of the strategy is that we may have on the computer many files of the same molecular information but in a different format. It requires some persistence on the part of the user to keep track of the files and to delete the obsolete ones.

Structure entry is accomplished in several ways. Atomic coordinates can be obtained from the literature, from the Cambridge Crystallographic Database, or from crystallographic studies. If the coordinates are not known, CAMSEQ/M (ref. 25) or CAMSEQ (ref. 20) can be used to construct a crude model, which can then be refined. CAMSEQ/M permits the chemist to enter structures graphically from templates or a Dreiding model-builder. CAMSEQ accomplishes the same thing using connection table input with or without coordinates.

Structure refinement involves the minor modification of bond angles and distances to minimize molecular energy. It can be done with molecular or quantum mechanical methods. For molecular mechanics calculations we use the MMPI and MM2 programs (ref. 26). Quantum mechanical calculations require more computer time than do molecular mechanical ones but they are easier to use for the refinement of the structure of molecules for which parameters are not available for the molecular mechanics programs. MNDO (ref. 23) or MINDO/3 (ref. 4) are currently used to do such structure refinement. Since we anticipate the need to do more accurate quantum mechanical calculations, we have installed the GAUSS79 and HONDO (ref. 12) ab-initio programs.

Electrostatic calculations are performed by using one of the quantum mechanical programs, MINDO/3, (ref. 19),MNDO, or GAUSS79. These programs are used to calculate gross atomic populations which are input to the program VSS. The VSS program (ref. 8) calculates the electrostatic potential energy due to the interaction of a unit positive charge with the atomic charges calculated by the quantum mechanical programs. Contours of the potential energies give information about the site at which electrostatic interaction with a molecule is most favored energetically. These contour maps also help in comparing two molecules according to their charge distributions. (ref. 9)

Conformational analysis is typically done with CAMSEQ. It performs a rigid rotation and calculates energies with an empirical potential energy function. The energy is considered to be the sum of steric, electrostatic, torsional, and solvation energy contributions. CAMSEQ produces a file of energy values that are used as input into CONTUR. CONTUR is a program that uses the DISPLAA software to produce contour plots such as seen in Fig. 2. We are currently installing the VAX version of CAMSEQ, CHEMLAB.

Molecular graphic display is done currently at Abbott using the XRAY (ref. 6) or CAMSEQ/M and SPACEFILL (ref. 25) programs. XRAY runs on the DECsystem-10 and supports a variety of graphic devices. We have modified it to make the spacefilling drawings to use van der Waal's radii, to give the user the option of labelling only heteroatoms, and to show superimposed molecules with one dotted. CAMSEQ/M and SPACEFILL run on the TEKTRONIX 4054 microprocessor. The programs display molecules either as Dreiding

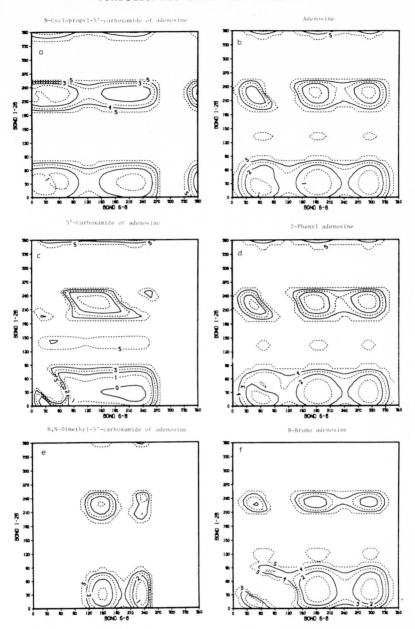

Fig. 2. Contour plots of the energy vs. clockwise rotation of the C8-N9-C1'-O1' torsion angle (Y-axis) and H4'-C4'-C5'-O5' torsion angle (X-axis). The plots are for (a) 5'-cyclopropyl amide, (b) adenosine, (c) 5'-amide, (d) 2-phenyl adenosine, (e) 5'-N,N-dimethyl amide, (f) 8-bromo adenosine, steric factors only.

back-bone structures or as space-filling models. The user may change the orientation, superimpose, and modify structures. Stereo-views of a structure can be prepared to simulate a 3-dimensional image.

A number of deficiencies exist within our current molecular graphic facilities. Plans are being made now to develop software for displaying molecular structures in color with features for depth-cuing, perspective, and real-time interaction.

ANALYSIS OF THE RELATIONSHIP BETWEEN AND WITHIN CHEMICAL AND BIOLOGICAL PROPERTIES OF MOLECULES

Multiple regression analysis is the standard tool used to investigate the linear free energy relationship between potency and a variety of molecular properties such as are collected in Table 1. For this purpose we generally use the SAS computer system for data analysis (ref. 1). This system is on our IBM-370. We prepare the data with a text editor on the PDP-10 and then submit it as a batch job through a high-speed link to the IBM. The output is sent back to a PDP-10 file to be edited or printed out.

Regression analysis of the first set of adenosine amides gave the following equation as the only statistically significant one:

$$\log(\frac{1}{C}) = 0.77 - 2.22\pi - 1.32\pi^2 + 1.47\ I \qquad (1)$$

$$R^2 = 0.86, \qquad n = 9, \qquad s = 0.43$$

In Eq. 1 we see a parabolic dependence of the logarithm of relative potency log(1/C) on the hydrophobic substituent constant π. The descriptor I is an indicator variable that is given a value of 1.0 for cyclic substituents and 0.0 for noncyclic ones. The equation shows that cyclic compounds are 1.47 log units more potent than their corresponding isohydrophobic analogues.

Equation 1 correctly predicted the potency of an additional seven analogues as shown in Figure 3. Obviously, such predictions of potency before synthesis is the objective of our work.

However, closer examination of the structure-activity relationships of all 5'-substituted adenosines showed that Eq. 1 is not the optimum descriptor of the structure-activity relationships in this series. For example, no amide in which both

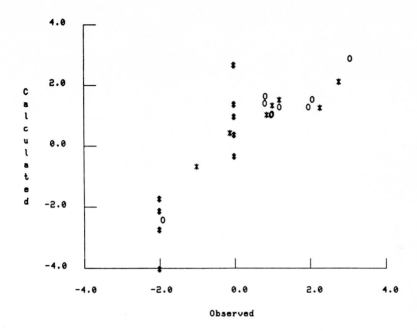

Fig. 3. Correct prediction of the potency of adenosine amides from
Eq. 1.

hydrogens have been substituted has any activity. Clearly
hydrophobicity alone cannot explain the inactivity of these
compounds.

Eigenvalue and principal component analysis have been shown to
be useful in the examination of the relationships between physical
properties in a set of analogues (ref. 16). For these calculations
we use the BMDP07 program (ref. 5). The data is prepared on the
PDP-10, sent to SAS on the IBM, and sent from SAS to BMDP.

The principal component analysis of the correlation matrix of
the physical properties tells one how many independent properties
there are in the matrix. In the case of true independent
modification of the value of each physical property, all
eigenvalues of the correlation matrix are equal to 1.0. On the
other hand, if there is high collinearity between the value of
some of the variables, the largest eigenvalue is larger than 1.0,
and the smallest ones are substantially less than 1.0. Another
parameter that is derived from the principal component analysis is

the squared multiple correlation of each variable with all other variables. This R^2 is small if the change in a variable is not correlated with the change in the value of the other variables. It approaches 1.0 as such changes become more and more highly correlated. In Table 1 we have listed the R^2 values for the physical properties in the set of adenosine amides. The rather high R^2 values suggests that the set was not well designed for regression analysis and that Eq. 1 may be somehow fortuitous: that the correlation with partition coefficient may be an artifact of the correlation of partition coefficient with other molecular properties.

Cluster analysis, that is calculation of the distance between compounds in property space, can be used to detect compounds that have essentially duplicate physical properties (ref. 16). Usually it is not necessary to synthesize such duplicates early in the exploration of a series, but if an interesting compound is found, synthesis of compounds close to it in property space may be indicated. We use the BDMP02M for these calculations.

Discriminant analysis is used to analyze structure-activity relationships for data sets in which the biological activity is of the active/inactive type rather than continuous. This type of analysis has been shown to have predictive value (ref. 15). For example, we were able to predict which 5'-adenosine analogues would lower blood pressure in the dog from a discriminant function: the compounds used to define the function were classified correctly 89% of the time whereas the new compounds were classified correctly 86% of the time.

Distance geometry calculations have been applied by others to the analysis of molecular shape as a determinant of activity vs inactivity (ref. 14) and to a QSAR type of analysis of binding potency (ref. 3).

Molecular graphics relies on human pattern recognition abilities to compare the complex three-dimensional properties of molecules. We are currently designing such a real-time color graphics display system. With the use of color a fourth dimension can be added to the molecular display. For example, the charge distribution in a molecule can be color coded on the computed van der Waals surface of a molecule. In one coding scheme, the charge distributions will be sorted in bins of increasing values. Each bin will then be colored to reflect the magnitude of the charge

(ref. 13). In this way hot spots in a molecule will be identified
and displayed, and these regions in different molecules
highlighted.

Such terms as molecular surface and hot spots cannot be
defined in a way that they can be handled with standard
statistical methods. While the calculation of the molecular
surface can be accomplished (ref. 13), mathematical comparisons of
several computed surfaces is difficult even for a sophisticated
image processing/artificial intelligence computer system. However,
the human eye is particularly well suited to pattern matching.
When the color coded three-dimensional image of several molecular
structures are presented to a chemist, similarities and
differences among the structures can be discerned in a way that no
current computer system can do. Such perceptions of similarities
and differences may then be used by the scientist to construct a
hypothetical map of the biomolecule that interacts with the
molecules being compared.

Geometrical modelling of complex molecular phenomena, such as
the unwinding of DNA, has also been accomplished by using
interactive three dimensional computer graphics (ref. 18).

The use of interactive graphics allows chemists not only to be
fed the results of structural studies, but also to feed back their
thoughts about possible molecular modifications. In this way
chemists can guide the computer in those areas in which it is most
deficient, namely creative thought.

FURTHER INVESTIGATION OF THE MOLECULAR PROPERTIES OF ADENOSINE ANALOGUES RESPONSIBLE FOR CORONARY SINUS VASODILATION

Since the regression equations that relate hydrophobicity to
potency of adenosines do not have universal applicancy, we
investigated this problem further. First, in plots we noted that
there appears to be a good negative correlation between $\log(1/C)$
and the length of the substituent, except that the $-CONH_2$ and
$-CONHMe$ analogues are less potent than expected. This is
summarized in Eq. 2:

$$\log(1/C) = 1.68 - 1.30 \ L - 2.70 \ S \qquad (2)$$

$$R^2 = 0.80, \ n = 16, \ s = 0.62.$$

In Eq. 2, the indicator variable S is set to 1.0 for the small substituents, methyl and hydrogen. The negative coefficient of S shows that the -H and -CH$_3$ analogues are 2.70 log units less potent than suggested by the correlation with length: this means that it is probably not shape alone but some more complex aspect of conformation that determines the relative potency. The negative slope of the term in L shows that larger substituents are also less active than the cyclopropyl analogue. There is no correlation of potency with either the minimum or maximum width of the substituents.

In molecules such as these, that is those in which there are several rotatable bonds, changing a substituent changes both the overall shape at any given combination of torsion angles (conformation) and also the probability that any particular conformation will be observed. Note that the substitution of a cyclopropyl group for a hydrogen atom or a methyl group increases potency by more than 100X. This dramatic increase is not easily explained by changes in physical properties such as increases in hydrophobicity or dispersion binding; accordingly, we have investigated the effect of substituents on conformation.

Selection of compounds for comparison is an important problem because, unlike when regression analysis is used, it is not technically feasible to include every compound, particularly those that have several rotatable bonds. We selected a sub-set of representative compounds to investigate the variation of conformation with structure. Since adenosine is the natural hormone, we included it. Presumably the conformation recognized by the adenosine receptor for coronary blood flow would be one not forbidden (within 10 kcal. of the minimum). We also investigated the 5' carboxamide and the 5'-N,N-dimethyl amide because the former is active whereas the latter is inactive. The cyclopropyl amide was included because it is the most potent known compound. The 2-chloro and the 8-bromo analogues were included because the former is as active as adenosine whereas the latter is inactive. The 2-phenyl and 5'-CH$_2$NH$_2$ analogues were also included because they represent different types of substitution and are slightly active, about the same as adenosine. For several of the compounds there is also some experimental information on their low-energy conformation.

Molecular geometry was calculated with the molecular mechanics
program MM2 from a starting structure built from the crystal
structure of adenosine and a Dreiding model-builder. Two ring
puckers were considered, 3'-endo and 2'-endo. The starting
structure for the 3'-endo was the crystal structure of adenosine;
the 2'-endo was generated by appropriate rotation about the bonds
of the ribose ring. In the geometry minimization the structure of
the heavy atoms of the adenine ring were held fixed. Since a
number of parameters had to be estimated, the force field was
verified by showing that it would give a structure close to the
X-ray structure of adenosine from crude coordinates.

Charges on the atoms were calculated with the CNDO/2 option of
CAMSEQ. Because of the lack of parameters, charges could not be
calculated for Br; we calculated the map for this compound using
no electrostatic term and also using the charges from the
corresponding chloro analogue.

Conformational properties were calculated with the potential
energy type function of CAMSEQ. Two energy terms were considered;
a 6-12 steric function and an electrostatic one. For the former
the parameters of (ref. 17) were used because on trial runs on
adenosine they produced the correct minima. For the electrostatic
function a dielectric constant of 3.5 was used.

Comparison of the analogues is straightforward. The critical
comparisons are shape (including the distance between the atoms
that bind to the receptor), and flexibility. In this case all
analogues have approximately the same bond angles and distances.
Accordingly, one can directly compare contour maps of energy as a
function of rotation angle. The various conformers are populated
according to the partition function, which is exponential in
energy (ref. 21). This means that the area encircles by any given
energy contour is a rough measure of the flexibility of the
molecule. The fraction of the total molecules in any one
conformation can be estimated by the fraction of the total contour
space that is accounted for by the conformation in question.

Our preliminary hypothesis is that the 2'-endo isomer is the
active one. PMR coupling constants suggest that the 2'-endo pucker
of the cyclopropylamide, by far the most potent analogue ever
tested, is the preferred one in DMSO solvent. In addition, our
calculations suggest that in this pucker the cyclopropyl analogue
is different in a subtle way from the unsubstituted analogue but

in a more dramatic way from the inactive dimethyl analogue.

Fig. 2 shows the contour plots of six analogues; the last two are inactive. The X-axis represents rotation about the H4'-C4'-C5'-O bond. The Y-axis represents rotation about the C8-N9-C1'-O bond. Fig. 1 shows the reference conformation. A comparison of the plot of the inactive N,N-dimethyl 5'-carboxamide with those of the active analogues suggests that it may be inactive because it cannot adopt conformations in which rotations about the C4'-C5' bond are between 0° and 100°. In other words, these results suggest that the active conformation is that in which the O5' is approximately 45° with respect to H4'.

Most of the maps show quite a bit of similarity between the rotation of N9-C1' at 30° (anti) and 210°(syn). There is no firm basis in these figures on which to decide if the syn or anti conformation is that which is bound by the receptor. However, since the unsubstituted analogue is active, we propose that the syn conformation is the active one because it appears more similar to the cyclopropyl analogue in this region of the map.

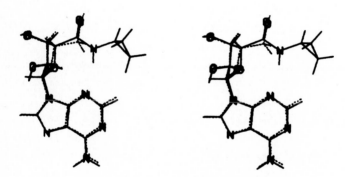

Fig. 4. Superposition of the minimum energy conformation of the 5'-cyclopropyl carboxamide of adenosine with the corresponding low energy conformation of adenosine.

Fig. 4 shows the superposition of the cyclopropyl carboxamide in its minimum energy conformation with adenosine in its corresponding low-energy conformation. The above hypothesis does not explain the inactivity of the 8-bromo analogue. With only one analogue one cannot be sure if the inactivity is due to steric repulsion by the receptor of the added group or to an unfavorable electrostatic change.

Further studies are now in progress to extend these calculations to other analogues. The problems with interpreting these results are typical with this type of analysis of structure-activity relationships. They also illustrate that if one wants to get conformational information from a structure-activity study, then the analogues should be designed from conformational considerations.

SUMMARY

We have described how a variety of computer applications can be helpful to a medicinal chemist in the design of new drug molecules. Special emphasis has been given to the use of linear free energy descriptors of molecules coupled with multivariate statistical analysis of the data and, on the other hand, to modern computation of three-dimensional molecular structure coupled with real-time color graphics to aid the analysis of structure-activity relationships.

TABLE 1. Physical Properties of 5'-substituted Adenosine Carboxamides

R2	σ^*	f	B2	B4	L	MR	log(1/C)
n-Hexyl	0.00	0.27	1.90	6.33	9.07	31.99	−1.90
CH_2CH_2O	0.40	−0.41	3.09		10.40	50.27	−1.00
Benzyl	0.22	−0.42	3.09	3.11	8.24	38.79	−0.11
H	0.49	−2.18	1.50	1.84	2.93	9.81	0.83
Methyl	0.00	−1.94	1.90	3.08	3.53	13.39	0.85
$(CH_2)2c$-pro	0.00	−0.19	2.24		6.76	29.93	0.88
c-Pentyl	0.00	−0.13	2.86	4.07	6.06	29.93	1.00
n-Propyl	0.00	−0.86	1.90	4.36	6.07	22.69	1.02
Propynyl	0.76	−2.12	1.90	2.99	6.62	15.87	1.03
Allyl	0.23	−1.29	1.90	4.82	5.31	20.63	1.22
$CH_2CHOHCH_3$	0.21	−1.81	1.90		6.07	24.21	1.22
i-Propyl	0.00	−0.99	1.90	4.15	4.96	22.69	2.00
Ethyl	0.00	−1.40	1.90	3.42	4.96	18.04	2.10
CH_2CH_2OH	0.21	−2.32	1.90	4.36	6.07	19.56	2.30
c-Butyl	0.00	−0.67	2.65		5.40	34.58	2.81
c-Propyl	0.00	−1.21	2.24	3.64	4.72	20.63	3.10
R^2	0.56	0.75	0.75		0.84	0.92	

REFERENCES

1 A.J.Barr, J.H. Goodnight, J.P. Sall, W.H. Blair, D.M. Chilko, "SAS User's Guide 1979 Edition". SAS Institute, Raleigh. (1979)

2 D.B.Boyd, D.K. Herron, W.W. Lunn, W.A. Spitzer, J. Am. Chem. 102 (1980) 1812.

3 G.M.Crippen, J. Med. Chem., 24 (1981) 198.

4 M.J.Dewar, J. Am. Chem. Soc., 97 (1975) 1285.

5 W.J.Dixon, M.B. Brown, "BMDP-77 Biomedical Computer Programs P-Series". Univ. Calif., Berkeley. (1977).

6 R.J.Feldmann, S.J. Heller, C. Bacon, J. Chem. Doc.,12 (1972) 234.

7 T.Fujita, J. Iwasa, C. Hansch, J. Am. Chem. Soc., 86 (1964) 5175.

8 C.Geissner-Pretre, A. Pullman, Theor. Chim. Acta, 25 (1972) 83.

9 P. Gund, J.D. Andose, J.B. Rhodes, G.M. Smith, Science, 208 (1980) 1425.

10 C. Hansch in "Biological Activity and Chemical Structure" (J.A. Keverling Buisman, Ed.) Elsevier, Amsterdam. 47 (1977).

11 C. Hansch, A. Leo "Substituent Constants for Correlation Analysis in Chemistry and Biology, Wiley, N.Y. (1979).

12 H. King, M. Dupuis, J. Rys, "NRCC Software Catalog" Prog. QH02 (HONDO5), 1 (1980) 61.

13 R. Langridge, T.E. Ferrin, I.D. Kuntz, M.L. Connolly, Science, 211 (1981) 661.

14 G.R. Marshall, C. Barry, H.E. Bosshard, R.A. Dammkoehler, D.A. Dunn, in "Computer-Assisted Drug Design" (E.C. Olson and R.E. Christoffersen, Eds.), American Chemical Society, Washington. 205 (1979).

15 Y.C. Martin in "Strategy in Drug Research" (J.A.K. Buisman,Ed.) Elsevier, Amsterdam. 4 (1982) 269.

16 Y.C. Martin, J. Med. Chem., 22 (1979) 784.

17 F.A. Momany, R.F. McGuire, A.W. Burgess, H.A. Scheraga, J. Phys. Chem., 79, (1975) 2361.

18 T.J. O'Donnell, Comp. Graph., 15 (1981) 133.

19 J.A. Pople, D.L. Beveridge, Approximate Molecular Orbital Theory, McGraw-Hill, N.Y., (1970).

20 R. Potenzone, Comput. Chem. 13 (1977) 187.

21 W.G. Richards "Quantum Pharmacology", Butterworths, Boston. (1977).

22 H.H. Stein, R.N. Prasad, P. Somani, Ann. New York Acad. Sci.,
   255 (1975) 380.
23 W. Thiel, M.J. Dewar, QCPE Program, 428 (1981).
24 A. Verloop, W. Hoogenstraaten, J. Tipker in "Drug Design" (E.J.
   Ariens, ED.) Academic, N.Y. 7 (1976) 165.
25 H.J. Weintraub in "Computer-Assisted Drug Design" (E.C. Olson
   and R.E. Christoffersen, Eds.). American Chemical Society,
   Washington. (1979) 353.
26 Allinger, MMPI and MM2 programs. (1976).

ABSTRACTS from the Poster Sessions

1. AN EXTENDED VERSION OF CHEMICS FOR STRUCTURE ELUCIDATION OF ORGANIC COMPOUNDS CONTAINING C, H, O, N, S, AND HALOGENS

Hidetsuqu Abe, Shin-ichi Sasaki, Iwao Fujiwara, Tohru Okuyama and Takashi Nishimura
Toyohashi Universiy of Technology, Tempaku, Toyohashi 440 JAPAN

As mentioned in the preliminary lecture by one of the authors (S.S.), we have been developing the computer program system for structure elucidation, named CHEMICS. Now the system has been extended to be able to handle the compounds with C, H, O, N, S and halogens. The extended CHEMICS is substantially organized with the similar strategy to its predecessor.

At present, a new set of 'components' (a sort of substructures used throughout the system) has been settled and the preliminary version of extended CHEMICS has been developed, although it requires some refinement and modifications to maintain the confidence level shown by the predecessor and to make it practical.

The latest results obtained by using extended version will be presented.

2. RECURSION TECHNIQUES TO STUDY SURFACE AND END ELECTRONIC STRUCTURES

G. Biczo
Central Research Institute of Chemistry, Hungarian Academy of Sciences, H-1525 Budapest 114, P.O. Box 17, HUNGARY

A direct (ref. 1) and an indirect (ref. 2) LCAO recursion method (DRM ref. 1 and IRM ref. 2) was compared in (ref. 3). Such approaches are briefly reviewed in a comment (ref. 4) on a transfer matrix scheme applied for surfaceband calculations by Lee

and Joannopoulos (ref. 5), proposing the use of an inversion-free
DRM for perfect crystal surfaces (ref. 6). The essential identity
of several differently called IRM-s and their relations to the
DRM-s are shown in (ref. 7). We generalize here the inversion-free
DRM (ref. 6) for crystals with imperfect surfaces or interfaces.
Its computational advantages are emphasized when QZ^8 or LZ^9
algorithms are applied to solve its auxiliary equation ^{1,3,6}
which, at the same time, is the fundamental equation of the IRM-s
^{2,3,7}.

1 G.Biczo, Proc. 7th Int. Vacuum Congr. and 3rd Int. Conf. on
  Solid Surfaces, ed. by R. Dobrozemsky et al., (Berger and Sohne,
  Vienna, 177), Vol. I, pp. 407-410.
2 O. Fromm and J. Koutecky, in Catalysis in Chemistry and
  Biochemistry. Theory and Experiment, ed. by B. Pullman (Reidel,
  Dordrecht, Holland (1979), pp. 335-346.
3 G. Biczo, in Recent Developments in Condensed Matter Physics,
  ed. by J.T. Devreese et al., (Plenum, New York) Vol. 2 (1981)
  pp. 341-352.
4 G. Biczo, O. Fromm and J. Koutecky, "Comments on (Simple Scheme
  for Surface-Band Calculations. I-II.) Competition between Green
  Function and Transfer Matrix Methods", Phys. Rev. B, submitted.
5 D.H. Lee and J.D. Joannopoulos, Phys. Rev. B23 (1981) 4988 and
  4997.
6 G. Biczo, J. Koutecky and Anna Lee, "Inversion-Free Formulation
  of the Direct Recursion (Transfer Matrix) Method", Phys. Rev. B,
  submitted.
7 M. Tomasek and G. Biczo, Interrelations of the Mathematical
  Approaches Applied for Studying Surface Electronic Structures",
  in preparation.
8 C.B. Moler and G.W. Stewart, SIAM J. Number. Anal. 10 (1973)
  S241.
9 L. Kaufman, SIAM J. Number. Anal. 11 (1974) S997.

3. CONFORMATIONAL EVALUATION OF CYCLIC SYSTEMS VIA
MICROCOMPUTER

P. J. De Clercq
State University of Ghent, Laboratory for Organic Synthesis,
Krijgslaan, 271 (S4), B-9000 Gent, BELGIUM

A program, designed for microcomputer operation, is presented
for the semi-quantitative conformational evaluation of cyclic
systems starting from a two-dimensional structural diagram. The
central part performs an evaluation of the torsion constraint in
the analyzed cycle (cf. P.J. De Clercq, Tetrahedron, 37 (1981)
4277) and can be used independently. For each ringsize (presently
five-, six- and sevenmembered carbocycles) a specific program then
derives the set of geometrically allowed basic conformations with
inclusion of conformational energies (cf. P.J. De Clercq, J. Org.
Chem. 46 (1981) 667). For each derived form the user is provided
with a full qualitative description, with the relative strain
energy of the form, with the number of syndiaxial interactions and
with torsion angle values, that are calculated on the base of an
appropriate mathematical model. Using these data the eventual
prediction of the preferred conformation is made by the user. As
for now, in polycyclic systems, each cycle has to be examined
separately. Input information is given on an easy interactive
basis (menu and yes/no answers). The program is written in BASIC
(16K Extended Cromenco version) and, by appropriate chaining via
floppy-disk, never requires more than 18K in central memory. When
running on a Cromenco Z-2 system a full analysis, including input
and program chaining, only requires a few minutes. The program is
especially useful for the examination of flexible systems (e.g.,
sevenmembered rings) and should be a valuable tool, both for rapid
every-day conformational evaluation (i.e., in combination with
framework molecular model examination) and for generating trial
geometries for further energy minimization.

4. TREE-STRUCTURED SPECTRAL LIBRARIES FOR MOLECULAR STRUCTURE ELUCIDATION

Michael F. Delaney
Department of Chemistry, Boston University, Boston, Massachusetts 02215

Tree-structured information provides tremendous increase in access efficiency, compared to sequential storage. The construction of an information tree using the formal approach is in most cases prohibitively long. Recently Zupan (ref. 1) developed an approach which allows such a tree to be generated in an efficient manner. While the Zupan tree is not identical to the formal tree, we demonstrated how to efficiently convert a Zupan tree to the formal tree (ref. 2). In this presentation, we will discuss some of our concerns, conclusions and progress related to the use of tree-structured spectral libraries for compound identification.

The following topics will be discussed:

1 Can a heuristic tree search provide hit-lists identical to those obtained by sequential search?

2 How large a gain in search efficiency is anticipated from tree search for a real spectral library?

3 To what extent will experimental variation in unknown spectra cause the wrong path in a tree to be pursued?

4 A demonstration of the Zupan to formal tree conversion, using a tree distance metric to show progress.

5 A demonstration of programs which plot trees.

Our research is primarily focused on the problem of identification of compounds separated by gas chromatography using vapor phase infrared spectrometry (VPIR) and combined infrared and mass spectrometry.

1 Zupan, ANAL. CHIM. ACTA, 122 (1980) 337.
2 M.F. Delaney, ANAL. CHEM., 53 (1981) 2354

5. AUTOMATIC DEDUCTIVE SYSTEMS FOR CHEMISTRY

Paul A. D. de Maine
Electrical   Engineering   Department,   Dunstan   Hall,   Auburn
University, Alabama 36849.

An overview is given for CRAMS, the Chemical Reaction Analysis and Modeling System. This computer software system is designed to process data for reaction models with any combination of rate and equilibrium reactions.

Its essential characteristics are:

1 It is easy to use   because   the   interface supports a chemistry-like language that is very similar to the conventional "shorthand" that is used by chemists.

2 Within the available core-memory there are no restrictions on either the complexity or size of the reaction model.

3 It has a PREDICTIVE facility that can be used to determine what data are needed.

4 No mathematical or chemical assumptions are made. Data collected on disparate time scales can be used, and there is an automatic Error Detection and Corrective Action (EDCA) system that assures the accuracy of numerical answers.

5 There are pre- and post- processing facilities that permit entry of raw data and displays of selected information in a variety of ways.

6. COMPUTER-AIDED STRUCTURE ELUCIDATION FROM LOW-RESOLUTION MASS SPECTRA

B. G. Derendyaev
Novosibirsk Institute of Organic Chemistry, Siberian Division of the USSR Academy of Sciences, Prospekt akademika Lavrentyeva,9, 630090 Novosibirsk USSR

A computerized search system which employs the data on the masses and relative abundances of spectral peaks and primary neutral losses is designed for computer elucidation of chemical structures. Testing the system for more than 100 "unknowns", for which the computer catalog contained no spectra, has shown that the probability of recognizing a large structural fragment lies in the interval 60-80%, depending on the fragment size (100-50% of molecular weight). Reliability of such structural conclusion is 98%.

Recognition of structural fragments is based on the computer analysis of structures of reference compounds selected as best matches to the mass spectrum of the compound under investigation. The approach used for such recognition enables handling of reference structural information by representing it in different forms; original atomic-fragment code, and its modified atomic and skeleton codes.

7. A MOLECULAR ORBITAL PROGRAM PACKAGE FOR UNDERGRADUATES

Paul A. Dobosh
Mount Holyoke College, South Hadley, Massachusetts 01075

A package of molecular orbital programs has been prepared for use in undergraduate courses. The programs are based on the Extended Huckel (EH) and Intermediate Neglect of Differential Overlap (INDO) theories. They are designed to use simple and compatible input and to allow reasonable control over output. The coding is compact and very linear to allow students to follow the logic of the programs and to encourage student researchers to modify the programs.

The two theories are represented by four separate programs

which use many of the same subroutines and make uniform use of variables. The programs are:

EXTHUC — non-iterative extended Huckel program which allows complete control over atomic orbital parameters.

FRGMNT — a fragment analysis program based on extended Huckel theory. A molecule may be broken into an arbitrary number of fragments and the molecular orbitals for the molecule may be expressed in terms of fragment molecular orbitals. This method has seen much use by Hoffman in the analysis of the conformations of organometallic compounds.

INDOCL — a self-consistent field, single-determinant, Hartree-Fock program based on the INDO theory and suitable for closed-shell molecules containing the elements hydrogen through fluorine.

INDOPN — an unrestricted Hartree-Fock version of INDO for open-shell molecules, e.g. radicals and triplets. The program also calculates Electron Spin Resonance hyperfine coupling constants.

In addition to the main programs, two peripheral programs are included, one a molecular orbital exercise to illustrate the variational principle, the other a program to compute atomic coordinates for input to the main programs.

The programs will be distributed, along with instructions and examples of use, through the Quantum Chemistry Program Exchange. It is anticipated that more programs will be added to the package periodically.

8. THE USE OF ORTHOGONAL TRANSFORMATION IN THE CHEMICAL MEASUREMENTS

L. Domokos
Budapest Technical University, Institute for General and Analytical Chemistry, HUNGARY

The modern analytical chemical measurement and signal processing techniques need always more and more mathematical and computational aids. Many of these methods are directly or indirectly based on some kind of orthogonal transformation, like Fourier, Walsh, Hadamard, principal components, etc.

The poster is dealing with different types of orthogonal transformations and with their use in analytical measuring, signal enhancement, feature selection, pattern recognition etc., pointing out the common root of these very various applications. Computational and educational aspects are also emphasized.

9. NEW SOFTWARE DEVELOPMENTS IN A SPECTROSCOPIC LABORATORY

Charles L. Dumoulin and George C. Levy
N.I.H. Biotechnology Resource for Multi-Nuclei NMR and Data Processing, Departmet of Chemistry, Syracuse University, Syracuse, NEW YORK 13210

Several software projects are at various stages of completion at the N.I.H. Biotechnology Resource located at Syracuse University. These projects include a highspeed local area network (WARPATH), a powerful spectral data analysis package (ORACLE), and a molecular dynamics modeling program (MOLDYN). This poster will primarily present details about ORACLE and its application to spectroscopic data processing.

ORACLE is a large, modular software package used for the reduction and analysis of spectral data. In its current implementation, ORACLE supports extremely large data sets (1 million floating point data points). In addition, the user can exercise a wide range of control over various computational functions. For example, algorithms for the functions of peak identification and quantification baseline flattening and NMR T_1

analysis perform without user interaction in the default mode, but allow the user to specify various control parameters and limits as necessary. ORACLE also makes extensive use of color raster scan graphics displays. Because of its modular design, multiple terminal types are easily supported. Currently the software supports high speed Lexidata 3400 image processors and lower speed Tektronix 4027A color graphics terminals. ORACLE will be extended this summer to include support for two dimensional spectroscopy and FTIR specific functions.

10. EVAPORATION ANALYSIS CONTROLLED BY A MICROPROCESSOR

D. Fassler, H. Hobert, H. Wachtel
Department of Chemistry, Friedrich-Schiller University Jena, GDR

Evaporation analysis is a new method for the rapid characterization and quantitative determination of organic materials in different samples. Pure substances or mixtures, organic impurities in inorganic material, aqueous solutions or emulsions, surface films on metals or glasses can be investigated by this method. The measuring principle is simple: the sample is heated using a linear temperature programme. All organic components, which are evaporated or give gaseous products by decomposition are signalized by a flame ionization detector, the flame of which is burning in a short distance above the sample.

The apparatus for the evaporation analysis (patented, Pat. No. 130876,GDR)is combined with a microprocessor (K1520, Robotron). The processor is used to control the temperature programme. Several evaporation curves (a set of equidistant temperature-intensity-values) can be recorded and stored. A signal-averaging is possible and the difference between two curves can be computed. The results can be presented in a graphical form and also on paper-tape for an off-line treatment.

The whole system is very simple. Because the time of an experiment is ony 15 min., this combination is well suited for demonstrations and students exercises in the field of labor automation.

11. DYNAMIC SIMULATION OF SOLIDS AND LIQUIDS USING THE ICL DISTRIBUTED ARRAY PROCESSOR

David Fincham
Queen Mary College, University of LONDON

Dynamic simulation (or 'molecular dynamics') is a powerful method for the study of solids and, particularly, liquids. Usually a system of a few hundred molecules is simulated, and made pseudo-infinite by the application of periodic boundary conditions. The classical equations of motion of the molecules are solved in a series of time steps, at each of which the force and torque on each molecule due to its interaction with its neighbors must be evaluated. Usually polyatomic molecules are treated as rigid assemblies of interacting sites. When the system has reached thermal equilibrium it can be used to study the thermodynamic, structural and dynamic properties of the liquid, the essential input being the form of the assumed pair potential between interacting sites. Developments of the basic technique include non-equilibrium methods for the study of transport processes, and methods for performing simulations at constant pressure rather than constant volume.

Because of the double loop over pairs of interactions involved in the force calculating these simulations are very time consuming, and hence expensive, on conventional serial computers. The method is however very suitable for parallel computation, since all the pair interactions may be evaluated in parallel, and the motion of all molecules in the system may be integrated in parallel. The ICL Distributed Array Processor is such a parallel computer. It consists of an array of 64 x 64 single-bit processing elements each of which has its own 4096 bits of data memory. The processing elements execute the same instructions in parallel on their own data, and an activity register controls the assignment of the result. The whole device forms a specialized store module of an ICL 2900 series computer. It is programmed in DAP Fortran, a language which includes extensions to enable the handling of parallel operations to be expressed in a simple manner.

The DAP has proved to be very suitable for performing dynamic simulations, and systems as large as 4096 molecules have been studied. Current work in progress includes the study of orientational ordering of molecules absorbed on surfaces; the crystalline to plastic phase transition in SF_6 ; the formation of azeotropes in mixtures of carbon dioxide and ethane; and the properties of cyclopropane modelled by a three-centre Lennard-Jones potential.

12. A CONVENIENT COMPUTER APPROACH TO THE COMPLETE AND UNIQUE NAMING OF MOLECULES

Ph. Floersheim, H. R. Haegi and A. S. Dreiding
Department of Organic Chemistry, University of Zurich, SWITZERLAND

Based on the principles of a new nomenclature theory (ref. 1), a solution is offered to the problem of generating unique names (codes) for all types of molecules. The method is implemented in a computer program called ONOMA.

ONOMA conducts a conversation with the user, first leading him to a description of the molecule's constitution (connectivity and atom types). It then looks for the potentially rigid substructures (called centrons) which require further description. The centrons, on which information is stored in ONOMA, at present include square-planar, tetrahedral, pyramidal, trigonal-bipyramidal and octahedral arrangements, as well as such polycentrons as double bonds, allenes and biphenyls. For each centron ONOMA asks the appropriate questions, depending on preceding answers, to obtain the minimum information (angles, dihedral angles and orientations) which is sufficient for its description. When ONOMA has collected answers for all the centrons the description is complete.

The user can then ask for the canonization, which eliminates all the remaining arbitrariness of the description. Such arbitrariness is due to the chosen order of input and includes the numbering of the atoms. The unique description, obtained in this way, is taken as the structural name. ONOMA assures uniqueness of the structural name by always applying the same rules and the same canonization. It is not essential for the user to know the rules or to understand the canonization. The ONOMA name consists of a list of the atom types, the bonds as well as the non-redundant bond angles, dihedral angles and orientations, each of these structural aspects being represented by the atoms involved and each atom being represented by its canonical numbering.

The canonization algorithm is conceived so as to find equivalent atoms. On request ONOMA prints out the proper symmetry group of the molecule, be it rigid or not, and the stereotopic relationships of its atoms; it also states whether the molecule is chiral or not.

REFERENCES
1 K. Wirth, M. K. Huber, D. Pazis, F. Siegerist, Ph. Floersheim
   and A. S. Dreiding, to be published.

13. EVALUATION OF SEVERAL CONNECTED DATA MATRICES — PATH MODELLING

I. E. Frank, B. R. Kowalski

Several multivariate statistical methods have been developed to extract useful information from a data matrix describing one influence sector (mass spectra, chromatograms, concentration of pollutants from one sampling site, etc.). These methods however put all the measured variables into one matrix. Segregation of variables coming from different influence sectors is not possible.

Path modelling based on partial least squares is a new algorithm to handle several connected data matrices (blocks) representing connected influence sectors (several sampling sites on a river, structural and biological activity features of a molecule, etc.).

This paper describes the algorithm of path modelling, the structure of the program package PLS-2, interprets the information regarding the features, the connection among blocks and the prediction of dependent block variables. The connection between factor analysis, multiple regression, canonical correlation and path modelling is explained.

The method is demonstrated using a five block data set, analyzing the influence of run-off from strip mine spoils on the water quality of a river. Eleven water quality parameters were monitored at five sites over 25 months.

14. FACTOR ANALYSIS IN SPECTROSCOPY: IS THERE A SAVE AUTOMATIC DETERMINATION OF THE NUMBER OF FACTORS?

C. Jochum, B. R. Kowalski, I. Frank

Factor Analysis is probably the most popular multivariate statistical method used in chemical data analysis. Among its many applications are the evaluation of environmental data, spectroscopical data analysis and its use with hyphenated methods such as GC/MS and LC/UV.

Although there are many "automatic" or semi-automatic" techniques which assist in factor interpretation (e.g. target transformation, communality iteration, self modelling curve resolution), the determination of the correct number of underlying factors is in most cases still done "by hand", i.e. by using some heuristic criteria.

This paper describes two algorithms which allow the automatic determination of the number of factors in UV/VIS and MS-spectroscopical applications. They most likely will also work on other spectroscopical data. Both procedures are based on cross-validation (CR). The CR technique deletes part of the data of a data matrix, predicts these data from the remaining data of a data matrix, with a given number of factors and calculates the residual variance between predicted and deleted values.

In the first approach, an F-test with 99% confidence limit determines whether this residual variance is significant, i.e. the extracted factors do not fully represent the data. In this case, an additional factor is included and the whole procedure is repeated.

In the second approach the extracted factors of the partially deleted data set are compared with the original factors (undeleted data set). If the angle between the respective factors lies below a certain data noise level, the extracted factors are significant.

Both methods are compared and demonstrated on several GC/MS - and LC/UV - data sets.

15. PREDICTION OF CHEMICAL REACTIONS AND REACTIVITY

J. Gasteiger, B. Christoph, M.D. Guillen, M.G. Hutchings, M. Marsili, H. Saller, I. Suryanarayana
Institute of Organic Chemistry, Technical University Munich, D-8046 Garching, WEST GERMANY

The program system EROS (Elaboration of Reactions for Organic Synthesis) can be applied both to the prediction of chemical reactions (forward search) and to the design of organic syntheses (retrosynthetic search).(ref. 1) Reactions are treated as bond and electron shifting processes and are generated in a combinatorial manner. Thereby access is gained to known and novel reactions.

To extract the chemically feasible reactions from among the formally possible ones models and algorithms are being developed for automatically defining reaction site and for evaluating reactions. Thermochemical parameters allow the calculation of reaction enthalpies with high accuracy. (ref. 2) Methods have been developed for the rapid calculation of charge distributions, (ref. 3) electronegativities of molecular substructures (ref. 3) and effective substituent polarizabilities. (ref. 4) Calculations of and correlations with physical data (dipole moments, ESCA shifts, ^{1}H- and ^{13}C-NMR parameters) have shown the physical significance of these quantities. Studies with data on gas phase reactions demonstrated that the magnitude and the attenuation of inductive and polarizability effects on chemical reactivity can be quantified. Systems investigated include the proton affinity of alcohols, ethers, thiols, thioethers, amines, aldehydes, ketones, carboxylic acids, and esters. The same approach can be taken to calculate the acidities of alcohols, both in the gas phase and in solution, and substituent constants of LFER. Preliminary studies with biological data show the merits of the detailed electronic description of molecules by charges, electronegativities and local polarizabilities in QSAR.

REFERENCES

1 J. Gasteiger, C. Jochum, Topics Curr. Chem. 74 (1978) 93

2 Gasteiger, Comput. Chem. 2, (1978) 85; Tetrahedron 35 (1979) 1419

3 J. Gasteiger, M. Marsili, Tetrahedron 36 (1980) 3219; M. Marsili, J. Gasteiger, Croat. Chem. Acta 53 (1980) 601

4 J. Gasteiger, M.G. Hutchings, submitted

16. THE APPLICATION OF FACTOR ANALYSIS TO THE STUDY OF THE X-RAY INDUCED REDUCTION OF $KClO_4$

R. A. Gilbert
University South Florida, Tampa, Florida

Factor analysis has never been employed to follow any chemical reaction with the intent of determining the kinetics of the reaction. This presentation reviews the characteristics of factor analysis and then explores the use of FA to study the x-ray induced reduction of material to be analyzed by an x-ray photoelectron spectrometer. Such reductions are significant in surface analysis because they produce sample degradations and erroneous results. However, this reduction phenomenon does provide an illustration of FA's use in resolving the complex data that the reactions produce and provides kinetic information about the sample reduction. Results on the reduction of perchlorate and other samples are presented to illustrate FA's utility as a data analysis procedure.

17. DATA ANALYSIS USING STEREO GRAPHICS*

Stanley Grotch
Lawrence Livermore National Laboratory, Livermore, California

The techniques of multivariate statistics and pattern recognition are receiving widespread application in many problems of data analysis. The graphical representation of complex data continues to be of great interest. In many cases involving high dimensionality, transformations such as principal components

analysis or non-linear mapping can substantially reduce the number
of dimensions while still retaining the essential quality of the
original data. Frequently, a reduction to only three-dimensions
will   retain 90+% of the   variance of the   original data.
Nonetheless, the problem still remains of simply, yet effectively
showing these transformed data in a three-dimensional plot.

Several techniques for displaying either: (1) discrete points
(generally obtained from experimental measurements) and/or (2)
surfaces (generally derived from known or fit functionalities) in
three-dimensions are illustrated. A set of very general plotting
programs has been developed permitting the user to easily exercise
a variety of plotting options: projections, gridding, color, least
squares data fitting, class distinctions. The user may view the
plots in real time from any perspective and simply change most of
the program control parameters, producing new plots, iteratively.
One particularly informative presentation shows together in a
single plot: (1) the raw data points used to derive a given model,
(2) an open wire mesh surface of the derived model, and (3) the
contours of the model obtained.

With software editing, any combination of any rectangular
portions of these plots may be easily mapped onto a single frame,
thereby providing merged or overlaid plots. This approach is a
simple, yet very effective means for generating side-by-side
stereo pairs of data which may be seen in true stereo using a
variety of inexpensive viewing devices.

In most instances, presentation of three-dimensional data
using stereo graphics provides a more profound insight than does
the traditional monocular isometric plot. A series of color stereo
pairs panning around a scene is particularly informative. A number
of both color and black/white stereo examples taken from real
problems will illustrate these points.

*Work performed under the auspices of the U. S. Department of
Energy by the Lawrence Livermore National Laboratory under
contract number W-7405-ENG-48.

18. CHEMCALC, AN INTERACTIVE LABORATORY COMPUTER SYSTEM ON A
    PDP-11/23

C. F. Hammer & M. M. Kornbluth
Department of Chemistry, Georgetown University, Washington, DC
20057

CHEMCALC is a computer program for a PDP 11/23 that is an
amalgamation of a series of chemistry-related programs. By putting
these programs together, they can share many standard routines,
such as those concerning input-output and calculating the inverse
or determinant of a matrix. In addition, programs that utilize
elemental data (such as LABION & MOFO) read initial elemental data
from user-created disk files, and will ask the user for data on
any unknown elements. (The default data file consists on data for
C, H, O, N, S, Si, P, F, Cl, Br & I; information on other elements
can be added to the file or entered while the program is running.)
CHEMCALC currently consists of the following programs:

QUINS, Quantitative Ultraviolet, Infrared, Nuclear magnetic
    resonance and Mass Spectroscopies, developed by C. F. Hammer,
    R. B. Joseph, D. F. H. Swijter and M. Kornbluth, can
    calculate the concentrations of unknown mixtures via the
    usual Beer's law method, from known or calculated (via a
    least squares fit of absorbance vs. concentration)
    absorptivities, or by the derived matrix technique developed
    by Hammer & Joseph. This latter technique is a least squares
    treatment that computes an overlap contribution matrix that
    is an exact fit of one or more known mixtures. It improves a
    typical multicomponent quantitative technique by a factor of
    3 to 8 times.
EVMAT, MATrix EValuation, developed by C. F. Hammer, D. M. Joyce
    and M. Kornbluth, is used for the selection of the optimum
    matrix from over-determined spectral data, to be used in the
    quantitative spectroscopy program QUINS. EVMAT evaluates all
    possible unitary matrices derivable from the input
    rectangular matrix. It assumes the maximum independence of
    equations is the matrix having a determinant whose magnitude
    is nearest to 1.0, this being the series of data points
    having the least amount of spectral overlap.
LABION, LABelled ION mass spectrum simulation, developed by C.

F. Hammer, T. R. Reel, A. J. Vlietstra and M. Kornbluth, calculates the theoretical ion cluster intensities for any combination of elements and any % labelling. All or some of the atoms can be isotopically labelled. In addition, the overlap of the calculated spectrum can be compared with experimental spectra, including approximate comparisons with spectra with P-nH peaks.

MOFO, determination of MOlecular FOrmulae, developed by Hammer, Vlietstra, Sasaki, Ishida, Ochiai & Kornbluth, determines the molecular formulae of all possible compounds using MS data, PMR integration and C-13 NMR peak intensity data. It then ranks the possible formulae based on comparisons to the data entered. The correct formula is chosen 75% of the time with a medium resolution mass spectrum, about 95% of the time with MS & proton NMR, and very rarely misses with that data plus C-13 NMR data.

LABDET, LABelled DETermination & isotope analysis (currently part of MSMS on the CIS system), developed by C. F. Hammer, A. J. Vlietstra and M. Kornbluth, determines the mole fractions of isotopically labeled (H-2, C-13, N-15 or any P+1 AMU isotope) species from the mass spectra of the non-labeled and labeled compounds, taking into account all natural abundant species and P-H, P-H2 or P+H (associated isotope effects can also be included) contributions using a multiple least squares simultaneous solution treatment of the data. The contribution matrix is derived by the program based on requested input data, including any isotope effects that occur in the mass spectrometer. The % label (abundance) of the heavy isotope is then determined assuming an equilibrium labeled system.

The mole fractions of the individual labeled species are determined based on the mass spectrum of the labeled compound and a matrix derived from the MS of the non-labeled so that all P+H & P-H,H2 effects can be considered.

19.   A PROTOTYPE STANDARD REFERENCE DATA BASE SYSTEM FOR
MICROCOMPUTERS

Joseph Hilsenrath and John S. Gallagher
Office of Standard Reference Data, National Bureau of Standards,
Washington, D.C. 20234

The   versatility   and availability   of microcomputers in
departments of chemistry provides the Office of Standard Reference
Data as an opportunity to extend its reference data service to the
educational community. The NSRDS MICRODATA system discussed here
consists of a number of small data bases and associated programs
to   retrieve and display them. It has been programmed in Basic and
runs on the APPLE II (APPLE II is a registered trademark of APPLE
COMPUTER CO.) Computer. The data and programs fit on a single
diskette.

NSRDS Microdata System Characteristics

The system includes programs to access the disk-resident data
bases and perform a variety of operations: searching, computing,
displaying, printing, plotting, etc. Listed below are more
specific functions of the system.

1 Compute tables of themodynamic properties of steam and
isobutane over a wide range of temperature and density or
pressure.

2 Display and/or print, in narrative form, 38 electronic and
bulk properties for a given element.

3 Display and/or print any one or the   entire table of
Fundamental Physical Constants.

4 Browse through the data file, element by element.

5 Display and/or print the current table of atomic weights.

6 Plot one or more atomic properties as a function of atomic
number.

7 Create a subfile of data elements for use in specialized
user-generated programs.

20. COMPUTER   PROGRAM SYSTEM   FOR STRUCTURE   ELUCIDATION BY
     ARTIFICIAL INTELLIGENCE METHOD

Z. Hippe, B.   Debska,   J. Dulidan, B. Guzowska-Swider, J. Jaraczz
and Z. Szwajczak
Technical University, 35-959 Rzeszow, POLAND

A new extended version of the system for structure elucidation
of organic compounds was elaborated. The system SEAC-2, in
comparison to its pilot prototype (ref. 1) is programmed with
FORTRAN for large framework machine ODRA-1305 and provides more
flexibility   for   the   user.   The   6th   level   of   realization
incorporated in SEAC-2, allows to generate detailed chemical
structure with lowest possible machine time. Besides, the system
has   incorporated   more   powerful   routines   for   checking   the
correctness of identification thus less informational isomers are
being generated.

REFERENCES
1 B. Debska, J. Duliban, B. Guzowska-Swider, and Z. Hippe: Anal.
   Chim. Acta, Comp. Techn. & Optimiz., 133 (1981) 303

21.NOVEL PHILOSOPHY IN DISCOVERY OF SYNTHESES OF COMPLEX MOLECULES

Z. Hippe, G. Fic, R. Hippe and W. Szumilo
Technical University, 35-959 Rzeszow, POLAND

To the two known distinct approaches to computer assisted
discovery of organic syntheses (refs. 1,2), a third is added. This
novel solution   exploits the   idea of storing - instead of
reactions- real technologies, i.e. information about completed
synthetic trees of chemicals being produced recently in industry
(in Poland). Over 170 technologies with some 1300 compounds are
stored on magnetic disks. When for a target molecule synthetic
routes are generated, the computer develops tree of syntheses, but
from the very beginning an algorithm for searching of structural
similarity between any   subgoal (generated)   and compound(s)
(stored) belonging to real technologies, is used. It is assumed,
that this solution offers two important results: the synthetic

pathways, at least at their final parts, are more realistic, and besides, machine time is substantially reduced.

REFERENCES

1 E. J. Corey: Q. Rev. Chem. Soc., 25 (1971) 455

2 I. Ugi, J. Bauer, J. Brandt, J. Friedrich, J. Gasteiger, C. Jochum and W. Schubert: Angew. Chem. 91 (1979) 99

22. FACTOR ANALYSIS IN CHEMICAL EDUCATION & RESEARCH

Darryl G. Howery

Department of Chemistry, Brooklyn College of C.U.N.Y., Brooklyn, N.Y.

The marriage of multivariate mathematical methods to computers opened a new era of data analysis and nurtured the discipline of chemometrics. Factor analysis (FA), a method for studying matrices of data, is a major technique in chemometric applications and promises to play a useful role in chemical education. With FA, the number of factors needed to represent chemical data can be determined, data can be correlated, and physically significant models for chemical data can be developed. The relation of FA to multiple regression is discussed.

Factor analysis may apply to any type of chemical data which can be represented as a matrix. Based on the kinds of designees represented by the rows and columns of the data matrix, the information obtained from a factor analysis are outlined. Two broad applications are stressed. In analytical chemistry, the number, the identity, and, in certain cases, the concentrations of components in a multicomponent mixture can be determined from FA. In physical chemistry, the factors responsible for solute–solvent interactions can be tested with FA and complete models for data can be identified. Results from recent research on thin-layer chromatographic data are employed to illustrate the FA approach.

In chemical education, FA occupies a central role among the multivariate techniques of chemometrics, as evidenced at a recent symposium on chemometrics in chemical education (Amer. Chem. Soc., Fall Meeting, N.Y.,1981). A qualitative introduction to FA can be used to introduce students to the important ideas of chemometrics.

Documentation for a computer program designed to accompany "Factor Analysis in Chemistry" by Malinowski and Howery is employed to help undergraduate students carry out research with factor analysis.

23. THE SELECTIVE DI-SOURCE PROPAGATION ALGORITHM TO SEARCH THE PATH IN THE DIAMONDOID HYDROCARBON REARRANGEMENT REACTION

Takeshi Iizuka [a], Nobuhide Tanaka [b], Masaru Imai [b], Tadayoshi Kan [c], and Eiji Osawa [d]

[a] Dep. of Chem., Fac. of Educ., Gunma Univ., Maebashi, Gunma, JAPAN; [b] Comput. Cent. & [c] Dept. of Phys., Gakushin Univ., Mejiro, Tokyo, JAPAN; [d] Dep. of Chem., Fac. of Sci., Hokkaido Univ., Sapporo, Hokkaido, JAPAN

Daiamondoid hydrocarbons such as adamantane $C_{10}H_{16}$, diamantane $C_{14}H_{20}$, and triamantane $C_{18}H_{24}$ can be prepared from their isomeric hydrocarbons, catatysed by $AlBr_3$. Although these preparations are "one pot synthesis", their rearrangement path are multisteps. Therefore searching for these rearrangement paths are usually time-consuming and troublesome works, if not skillfully done, even though it were aided by computer. SDSPA is a method to search the reasonable rearrangement path by molecular forcefield calculation along the shortest path which is found by DSPA graph-theoretically, being assisted by computer. By applying the SDSPA, reasonable paths for the triamantane rearrangement reaction reported by McKervey and for the diamantane rearrangement reaction reported by Turecek are efficiently predicted. It would be almost impossible to predict the path by any other methods already reported. The development of a new predictional method as shown in this paper will surely assist chemical experiment to be done more efficiently.

24. THE APPLICATIONS OF FUZZY PROCESSINGS IN ORGANIC SUBSTRUCTURE ELUCIDATION OF INFRARED ABSORPTION SPECTROSCOPY

Y. Ishida, T. Ooshima (UBE Industries,Ltd., JAPAN)
S. Sasaki (Toyohashi Univ. of Technology, JAPAN)

A program adapted for a micro-computer and designed to assist the chemist in his interpretation of IR spectra is described. A computerized interpreter requires the correlation table which is based on empirical determinations of peak locations and/or intensities that are able to be correlated with specific chemical functionalities. The peak locations and intensities, however, are stated in the "fuzzy" expressions such as "nearly 1720 cm^{-1}" and "medium to strong absorption". Therefore, a chemist encounters difficulty in determining the particular locations and intensities for the computer.

A key feature of SIGMA-IR (Structural Information Generator of Membership function Analysis of IR spectra) is that the fuzzy processings are employed for the confirming the presence/absence of particular substructures. The program is to make the chemist aware of the possibilities for these functionalities in every case. The main function of the program is closely modeled on the method of solving the FUZZY INVERSE PROBLEM, and the method is employed by fitting the Gaussian curve for the membership function in order to calculate the distance from the center of characteristic absorption.

This program has been implemented on a Hewlett-Packard model 85-A in the BASIC language. The computer is interfased to a IR spectrometer and supported by a software package where the computer acquires and stores spectral data. The program can process cyclic and acyclic molecules with only C,H,O,N,S,P, and halogens, and molecular weights up to around 300 amu. The molecular formula or the atomic composition, if available, may be entered by the chemist. There are 159 substructures available to the process. As output, the program prints probabilities for each substructure to be assigned as a part of the molecule. This program runs as a standard job requiring 6K byte read/write memory and 74K byte auxiliary cassette MT for the correlation table file. By use of SIGMA -IR, the computational assistance has become a reality.

25. COMPUTERIZED STRUCTURE ELUCIDATION BY MS/MS

J.V. Johnson and R. A. Yost
Department of Chemistry, University of Florida, Gainesville,
Florida 32611

Of all the chemical techniques for elucidation of the structure of organic molecules, mass spectrometry (MS) is undoubtedly the most useful. It provides the single most valuable piece of structural information, the molecular weight, as well as the mass of the major fragments (or substructures) of the molecule. Computerized techniques for the interpretation of mass spectra have had limited success, however, due to the difficulty in extracting structural information from the spectral data.

Tandem mass spectrometry (MS/MS) provides an added dimension of mass spectral data for computerized structure elucidation (ref. 1,2). Each of the fragment ions (sub-structures) in the normal electron impact mass spectrum of a compound can be selected by the first mass analyzer, the selected ion can then be fragmented by collisionally activiated dissociation, and its mass spectrum obtained by scanning the second mass analyzer. Systematic MS/MS analysis of all the fragments of a molecule elucidates all the substructures which arise from a specific substructure, as well as all larger substructures which fragment to give the selected one. This three-dimensional "map" of the fragmentation pathways of the molecule should be particularly well suited for computerized interpretation.

We are investigating the combination of MS/MS data with the GENOA and CONGEN techniques available on the SUMEX-AIM system at Stanford University. The structure elucidation process involves: 1) obtaining the mass spectra of all fragment ions in the compound's mass spectrum; 2) identifying the corresponding substructures by computerized search of a library of known fragment ions; 3) input of these substructures to GENOA for generation of all consistent molecular structures.

The triple quadrapole MS/MS techniques will be explained (ref.3), together with the processes for obtaining fragmentation pathways, library searching the substructure library, and generating molecular structures with GENOA. Examples of the applications of this technique will be presented.

REFERENCES

1 R.A. Yost, C.G. Enke, Amer. Lab., 13(6),(1981) 88-95.

2 R.A. Yost, C.G. Enke, Org. Mass Spectrom.,16,(1981) 171-175.

3 R.A. Yost, C.G. Enke, Anal. Chem., 51,(1979) 1251A-1264A.

26. COMPUTER ASSISTED STUDIES OF MOLECULAR STRUCTURE AND BIOLOGICAL ACTIVITY

Peter C. Jurs

152 Davey Laboratory, Department of Chemistry, The Pennsylvania State University, University Park, PA 16802

Structure-activity relationships can be investigated for large sets of organic compounds using computer-assisted methods. The combination of chemical structure information handling, force-field molecular modelling, substructure searching, and related methods with pattern recognition and statistical analysis provide an approach to SAR studies. An interactive computer software system (ADAPT) has been implemented to automate this approach to SAR studies and to allow its application to large sets of organic compounds. The ADAPT system provides the user with the ability to input, store, and manipulate sets of organic compounds, to develop three-dimensional molecular models, to generate molecular structure descriptors (including fragment, substructural, environment, geometric, electronic, molecular connectivity, and others), to gather together sets of descriptors for analysis, and to analyze these data sets using statistical or pattern recognition methods.

A number of SAR studies have been pursued, including studies of pharmacological drugs, olfactory stimulants, genotoxic compounds, (mutagenic or carcinogenic compounds), and anti-tumor compounds. Results obtained in several investigations will be presented to illustrate the applicability of this methodology to SAR studies.

27. 'SPEKTREN'-A SPECTROSCOPIC INFORMATION SYSTEM

I. Koehler, C.W.v.d.Lieth, J.Mowitz, H. J. Opferkuch, M.Zippel
German Cancer Research Center, D 6900 Heidelberg, WEST GERMANY

The spectroscopic information system 'Spektren' in Heidelberg uses a databank with ca 2000 infrared, ca 4000 cnmr and ca 25000 mass spectra.

The   original infrared spectra, as they   come from the spectrometer, have a resolution of 1 point per wavenumber and are being stored as such on magnetic tape for possible further investigation.

For analytical purpose two reduced versions are resident in the masterfile: one version containing the peaks and intensities of the spectrum (wavenumber and transmission of the bands as derived by means of a peakfinder program) serves for spectra match. The other so called plot version (contains 1 point per 4 wavenumbers) the plot of which is congruent with that from the spectrometer.

For updating the physical data of spectra from the spectrometer (ir, cnmr, ms) as well as for input of spectra from the literature, several checking procedures are automatically active in order to avoid logical errors: the molmass, the molecular formula and the connectivity table are being checked against each other, and for cnmr spectra the multiplicity and chemical shift are controlled.

In a match of the spectrum of an unknown compound with the library, the library spectra are arranged in the order of decreasing similarity with the unknown. The matching algorythm uses the ratio of the number of coinciding peaks (within a selectible tolerance) and the total number of peaks in both spectra (as well as the ratio of intensities for infrared spectra).

The output of the match result can be either the graph of the structures of the best fits, or a short or comprehensive print of the fitting spectra, or for infrared spectra, a plot of the curves of the spectra with all peaks (lower traces) and the coinciding peaks (upper traces) together with the graphs of the chemical structures of the best fits.

28. APPLICATION OF MICROCOMPUTER IN CHEMICAL INFORMATION PROCESSING

Liang Xi-Yun, He Pu, Chen Jing-Feng
Institute of Chemistry, Academia Sinica,Beijing, CHINA

Most chemical information processing systems have been developed in the large computers. However, with the continuously decreasing cost and increasing capability of microcomputers recently, it makes more and more sense to develop systems with microcomputers to store and retrieve in a chemical data base of relatively small scale. We have developed such a micro-system capable of handling organic compounds of special interest. Organic pesticides are processed with the aim of correlating the chemical structure and the properties of these compounds as well as providing a method for the identifition of the compounds by gas chromatography retention parameters. The system is implemented on a Commodore Business Machine Inc. PET 2001-8 personal computer with extended memory of 12 KB and two diskette drives for a total storage of 340 KB. Structural information and physical properties of ca. 200 pesticides are packed into a record of 255 bytes each are stored randomly on one floppy disk as physical blocks using the direct access feature of the PET/DOS commands. Hashing method is used in allocating physical records are managed by a control program and is transparent to the user. The linkage between the user and the record manager is provided by the master key file which maps the entry numbers to physical track and sector numbers. The system uses simple command-like language to retrieve information and search through various names and gas chromatographic data to find the chemical structure and properties of the appropriate compound.

29. APPLICATION OF MICROCOMPUTER IN MASS SPECTROMETRY

Liang Xi-Yun, Zhan Mei-Yi, Chen Zhi-Feng, Liu Chin-Kun
Institute of Chemistry, Academia Sinica Beijing, CHINA

The usefulness of computer in the acquisition of mass spectrometric data computer-aided interpretation of mass spectra is well known to everyone. Library searching technique is the most commonly used method to extract useful structural information from the mass spectra of unknown compounds. We have established a model mass spectra search system based on a microcomputer. This system allowed us to create a micro-size library of N-nitrosamines, provided the facilities to add new spectra and edit existing ones.

It also allowed us to generate and examine search keys, and finally to search through the library for an unknown spectrum suspected to be in the same class.

The hardware configuration is very simple and comprised a CBM/PET-2001 microcomputer with 12 KB of memory and a PET-2024 dual mini-floppy disk sub-system (340KB).The spectral library which contained both the descriptive information and spectral data are stored on one floppy disk, one file for each spectrum. The monitoring programs and the various search-key files are stored on the other floppy disk. The operation of the search system is supervised by a control program called "MS-MONITOR", which handles the dialogue with the user and directs the requests to various functional modules. The modules are stored as individual programs on disk are overlaid on the calling program using the chaining capability provided by the PET monitor. The mass spectral data base consist of about 150 N-nitrosamines.

The search algorithm involves two steps: a pre-search using a key-word produced by the "SDI" method selects those reference spectra presumably of the same group as the unknown, and the selected spectra are matched against the unknown using either forward and reverse searching technique. Both the pre-selection and matching scheme can be easily changed by substituting the appropriate subroutines. The effectiveness of different search algorithm can be compared.

30. USING THE QCPE HOLDINGS IN CHEMICAL EDUCATION

K. Lipkowitz

The Quantum Chemistry Program Exchange is a valuable resource for educational needs in chemistry. Over 400 programs have been tested, are well documented and are available at a modest rate. The chemistry department at Indiana-Purdue University has used this resource in graduate and undergraduate curricula to provide extensive hands-on use of state-of-the-art computational programs. A detailed example of a course enhancement in Physical Organic Chemistry is presented.

31. AN MLAB STUDY OF AQUATIC STRUCTURE TOXICITY RELATIONSHIPS

Robert L. Lipnick,
U. S. Environmental Protection Agency, Washington, D.C. 20460

William J. Dunn, III,
University of Illinois at the Medical Center, Chicago, IL. 60680

A quantitative structure activity study was performed on the acute toxicity of organic chemicals to the fish golden orfe (Leuciscus idus melanotus), using test data from a recent publication of Juhnke and Ludemann. (ref. 1) Octanol/water partition coefficients (Log P) were obtained from measured values stored in the Pomona College Medicinal Chemistry Bank and the ISHOW database, or else calculated using the CLOGP computer program. The computer data analysis was accomplished using the Mathematical Modeling Laboratory (MLAB) component developed by the NIH Division of Computer Resources and Technology. The fish toxicity data for the majority of the chemicals studied (alcohols, ketones, ethers, halides and hydrocarbons) were well fitted to a Hansch equation in Log P similar to that obtained previously by Konemann for guppies, and Veith et al., for fathead minnows. The remaining chemicals are generally more toxic than the equation predicts, and therefore exhibit more specific effects which mask baseline narcosis. Additional linear free energy related parameters are being investigated to establish further possible correlation.

REFERENCE

1. Juhnke and D. Ludemann, Z. Wasser Abwasser Forsch., 11 (1979) 161-164.

32. TARGET FACTOR ANALYSIS WITH A MICROCOMPUTER

Prof. Edmund R. Malinowski
Dept. of Chem. & Chem. Eng., Stevens Inst. of Tech., Hoboken, N.J.
07030

TARGET is a set of target factor analysis (TFA) programs
designed for use with a microcomputer. It is designed not only to
teach TFA principles but also to provide research capabilities.
The set of programs interweave, provide a wide range of utility
and are stored on a single disk as follows: (1) MAKE DATA FILE,
(2) PRINT DATA MATRIX, (3) PRINT TARGET DATA, (4) EDITOR, (5)
PRETREATMENT,(6) EVA (eigenvalue analysis), (7) AFA (abstract
factor analysis, includes uniqueness tests), (8) TFA (target
factor analysis, includes individual target testing and performs
combination transformation, (9) KEY SET (locates key sets of
"typical" vectors and performs spectral isolation), and (10) MRA
(multiple regression analysis). Error analyses are performed at
each stage, thus providing information to judge the accuracy and
reliability of the results.

33. DRACO: AN INTERACTIVE SYSTEM FOR BOOLEAN REPRESENTATION AND MANIPULATION OF 3D MOLECULAR MODELS

Mario Marsili and Philip Floersheim
Organic Chemistry Institute, Universiy of Zurich, SWITZERLAND

A quantity like a "similarity index" of two molecules may be
of advantage in finding relations between the "shape" of drugs and
their activity. Tackling this unsolved problem developed an
approach based on a logical description of a v.d.Waals molecular
model in a computer, which we call Boolean Matter Code, BMC.
A given portion of physical space is represented by a tensor
$T(108,108,108)$ of rank 3 and 108 components per axis. Thus the
space delimited by T contains more than million subspaces, each
one of them characterized by an indeces tripel IKL. A scaled
v.d.Waals model of a molecule is located with its center of
gravity in the middle of T. All the subspaces residing inside the
v.d.Waals range of the model are labelled with an 1, the others

not containing molecular matter are labelled with a O. To gain a compact description of the distribution in space of molecular matter bit-addressing techniques are used. The x and y axes of T contain each 108 computer words (PDP-10), but the z axis has only three words consisting of 36 bits each. Thus, not the words but the bits are used to fill in the 1's and 0's. Core memory can be saved by a factor of 36 in this representation. After generation of BMC's superposition of two of them is posssible by least-squares matching of selected substructures of the two corresponding molecules. Using boolean operators addition and subtraction of the two BMC's can be achieved, generating a so-called SUPER-BMC and an EXCLUDED-BMC respectively. Similarly, those subspaces shared in common by both BMC's are defined as COMMON-BMC. The ratio between COMMON and SUPER BMC stays for a degree of spatial similarity of the molecular models. Visualization of BMC's is feasible on our GT-41 graphics terminal. The subspaces inside the molecular model appear as lit points around the molecular skeleton. Molecular tomography is used in addition for display of particular cuts through the superposed ensemble for detailed inspection of structural features. For easy handling of the models auxiliary functions are available on our system: the molecular coordinates can be generated from force-field calculations on a rough model drawn on the screen with light-pen. Alternatively they can be recalled from data banks. Generation of different conformations, reflection throgh planes, rotation in space and many others are implemented as well.

34. MOLECULAR SPECTROSCOPY DATA BANKS IN STRUCTURE ELUCIDATION

I. F. Mikhailova
Novosibirsk Institute of Organic Chemistry, Siberian Division of the USSR Academy of Sciences, Prospekt akademika Lavrentyeva, 9, 630090, Novosibirsk, USSR

Data banks on various types of molecular spectroscopy allow both identification of known compounds and elucidation of the structure of unknown ones. A new approach is discussed which makes use of the fact that compounds selected by a computer as best matches to the spectrum of compound in question generally appear to be its close structural analogues.

In general, that approach provides a researcher with more structural data than the traditional interpretation of spectra on the basis of known spectrum-structure correlations. This is especially evident in the case of mass-spectra and the electronic absorption spectra. A set of structural data obtained by this method from different molecular spectra (IR, UV, MS, ^{1}H and ^{13}C HMR spectra) is normally complete enough for the investigator to make the appropriate structural conclusion.

Examples of solutions of structural problems in various fields of science are given.

35. USE OF PATTERN RECOGNITION FOR THE STRUCTURE-TASTE STUDIES ON L-ASPARTYLDIPEPTIDES

Yoshikatsu Miyashita, Yoshimasa Takahashi, Chiyozo Takayama, Kazuo Sumi, Hidetsugu Abe, and Shin-ichi Sasaki
Toyohashi University of Technology, Toyohashi, Aichi 440, JAPAN

It is known that the taste of L-aspartyldipeptides (L-Asp-NH-R) varies from sweet to tasteless or bitter with the chemical structure of their C-terminal amino acid moieties R. Pattern recognition techniques were applied to the structure-taste studies on L-aspartyl-dpeptides. According to the molecular theory of taste, the taste response is affected by the shape, size, and functionality of a molecule. In the structural representation of the compounds, these factors were considered. The shape and size factors of the substituents R were described by the STERIMOL parameters and molar refractivity, respectively. The Taft's O* was used to describe the functionality of R. The SIMCA method was applied to investigate the structure-taste correlation of dipeptides. The dimensionality of the sweet class model was determined by the cross-validation technique. The following significant conclusion was derived from the classification results by SIMCA and the display methods (nonlinear mapping plot). The structural requirements for the sweet taste of dipeptides are considerably different from those for the bitter one.

One of the authors (Y.M.) gratefully acknowledges the financial support of the Japan Association of Chemistry for the presentation of this research.

36. A MICROCOMPUTER-INTERFACED STOPPED-FLOW KINETICS APPARATUS WITH INTERACTIVE GRAPHICS

John W. Moore
Department of Chemistry, University of Wisconsin, Madison, WI 53706 and Kenneth W. Hicks, John Vidolich, Sharon T. Pittenger, Kalle Gehring and Robert G. Williams, Department of Chemistry, Eastern Michigan Univ. Ypsilanti, MI, 48197

We have interfaced an Aminco-Morrow stopped-flow apparatus to an S-100 bus microcomputer system that includes interactive video graphics and optional hardcopy graphics. Data collection and data analysis are automated and user interaction with both processes is facilitated by the interactive graphics. The microcomputer hardware consists of an S-100 bus mainframe with power supply, a Z80 processor, 64 K static RAM, dual floppy disk drives with controller, serial and parallel I/O, a console CRT terminal, a graphics interface with video monitor and light pen, a Tecmar A/D converter board, a dot-matrix printer,, and a digital plotter. The user is directed by prompts that appear on the console CRT. Data being collected or analyzed is displayed on the video graphics monitor. Console commands and the light pen are used to select portions of the data for analysis according to first- or second-order rate equations, and goodness of fit is displayed graphically as well as statistically. The system makes computer/instrument/human interactions extremely cordial to the user and thereby more effective.

37. SUBSTRUCTURE SEARCH ON A HIERARCHIC TREE OF CHEMICAL GRAPHS

Z.M. Nagy, T. Veszpremi[*], G. Csonka[*], P. Bruck[**]
[*]Dept. of Inorganic Chemistry, Budapest Technical University, [**]Central Research Institute for Chemistry of the Hungarian Academy of Sciences, 1525 Budapest Pf 17 VI 11, HUNGARY

The retrieval of a substructure is closely related to the problematics of subgraph-isomorphism. Since the latter problem is mathematically NP-complete, in case of sequential comparison of subgraphs the time requirement of the retrieval would grow much faster, than the number of molecules in the data base.

The Hierarchic Tree Substructure Search (HTSS) system avoids
this difficulty executing the exponentially increasing part of the
operations only once - during the generation of the data base. The
critical point of the approach is to store the information,
resulted by this preprocessing of the chemical graphs, in a
fashion, which facilitates quick and complete response for a broad
range of structural problems, while its mass storage requirement
grows slowly with the number and size of molecules in the data
base.

This problem is solved in the HTSS system generating a single,
hierarchic, rooted decision tree from the connectivity tables of
the molecules utilizing a gradually refined classification of
their atoms.

As a starting point, the whole database of structures is
handled as a single (but non-connected) graph; each molecule is a
connected subgraph of this large graph.

The classification of an atom begins with the definition of
its local characteristics ("colour"), like the number of its
neighbours, the type of the atom, the type of its bonds and size
of its rings.

Next, the colour of the atom is modified, taking into account
the colour of its nearest neighbours first, and extending this
procedure gradually, until the effect of the furthermost atoms has
been considered, or the classification is complete.

The differentiation, resulted by the gradual colourization of
atoms, previously belonging to a single class, is always reflected
in a branching of the decision tree.

The structural information, contained by such a tree, is
equivalent to the connectivity matrices of all molecules in the
data base, but common structural entities are stored together, and
only once. However, the hierarchic, rooted structure of the tree
facilitates the structure search to a large extent.

To find a given (sub)structure in the data base, the graph of
the structure is gradually coloured (as in the case of the tree),
and after each step, the path, spanned by the atoms, having the
same colour in the decision tree of the data base, is followed,
until a leaf is reached, i.e. the same (sub)structure has been
found in the data base.

If no leaf could be reached, our standpoint in the tree corresponds to the most similar substructure to be found in the data base.

This system has been implemented on a PDP 11/34 computer in Fortran-IV-Plus language, using a model data base of 1500 compounds.

38. PARTIAL STRUCTURE ANALYSIS BY MEANS OF SPECTRAL DATA EVALUATION

Hans J. Opferkuch, I. Koehler, C.W.v.d. Lieth,
J. Mowitz, M. Zippel
German Cancer Research Center, D 6900 Heidelberg, GERMANY

File search using different spectroscopical data bases as MS, IR, and C-NMR is a powerful tool in structure elucidation (evaluation of identity or similarity) of organic compounds.

In case of similarity relevant partial structures can be evaluated if an appropriate structure representation is present in the data base.

Starting with the resulting list of similar compounds produced by comparing and evaluating spectral data the partial structure, stored as HOSE codes together with the spectral data in the data base, of the corresponding hits were analysed. The histogram of such a partial structure distribution study shows two clearly separated classes of partial structures. While a lot of them are only 1-3 times present (statistically distributed) occur a very small set of these partial structures much more frequently. These are the relevant partial structures. Using the spectral data and structural representation of a known compound as "unknown" in the file search. This interpretation is proved by marking the relevant partial structures. This approach using C-NMR data where a direct correlation between partial structure and spectroscopic information exists is not restricted to this spectroscopic method. Using IR- or MS-data similar results were achieved.

39. METHODS OF HANDLING CHEMICAL STRUCTURE INFORMATION AS APPLIED TO SOME PROBLEMS OF ORGANIC CHEMISTRY

V. N. Piottukh-Peletsky
Scientific Information Centre for Molecular Spectroscopy,
Institute of Organic Chemistry, Siberian Division of the USSR
Academy of Sciences 630090, Novosibirsk (USSR)

A software for computer manipulation of chemical structure
information developed at the Scientific Information Centre for
Molecular Spectroscopy is described. This software includes the
chemical structure data base formation package, substructure
search programs and programs evaluating the degree of similarity
of graphs corresponding to chemical structures. Some problems of
canonical representation of chemical structures and of taking
into account graph- and space symmetry of molecules are discussed.
Special attention is paid to representation of nonclassical
chemical structures. These problems lie on the borderline between
chemical informatics and the chemical structure theory. Examples
in the field of structure elucidation by means of molecular
spectroscopy and revealing structure-spectrum relationships
illustrate applications of the structure manipulation system.

40. MOLECULAR SPECTROSCOPY DATA COLLECTION AND PUBLISHING SYSTEM

M.I. Podgornaya
Scientific Information Centre for Molecular Spectroscopy,
Institute of Organic Chemistry, Siberian Branch of the USSR
Academy of Sciences.

A specialized publishing system is described for collection of
spectral information, transferring spectral data to computer
storage and graphical output of computer generated forms for
publication.
The system automatically processes texts, chemical structures
and spectra of chemical compounds. The system involves some
specialized hardware including graphical input devices,
alphanumeric displays, plotters and computer output on microfilm.
Computer software provides control and editing of the initial

information, space planning and automatic scale selection for computer output publication forms.

The system is designed for the preparation of molecular spectroscopy publications in the field of mass spectrometry, ^{13}C-NMR spectroscopy and ^{1}H-NMR spectroscopy literature data indexes.

41. VERSATILE INSTRUMENT/MINICOMPUTER CONTROL SYSTEM FOR ANALYSIS IN MASS SPECTROMETRY

L. Radermacher

Zentralabteilung fur Chemische Analysen (ZCH), Kernforschungsanlage Julich GmbH, P.O. Box 1913, D-5170 Julich 1, WEST-GERMANY

A versatile instrument/minicomputer control system is described (ref. 1), which is mainly adapted for analysis in mass spectrometry. A flexible instrument control is accomplished by assembler written subfunctions and the choice of clearly arranged parameters in a high level language "CALL"-command (BASIC or FORTRAN). The computer hardware consists of a DIGITAL EQUIPMENT (DEC) PDP 11/34 computer with 28 k memory. The system disk is a floppy or hard disk (DEC RX 01/02 or RL 01/02). There are a main (LA 36) and an alternate terminal (VT 100), a line printer, and up to 6 remote terminals, if necessary. A computer interface provides with the three device addresses for input, output, and address data to the instrument interface assembly. The "CALL"-subfunctions accomplish a specific software address to select the desired control and monitor functions. These functions include: sample stage position, scaler-timer, magnet field, ratio plates, sample voltage, beam deflection, display for scope/recorder, sample current counter integrator, ion and beam current. The "CALL"-commands are part of a stand-alone extended BASIC operating system which contains numerous conveniences and performance improvements compared to DEC RT-11 BASIC. They refer to I/O capability, faster disk access and program execution, and multiple drive file searches. Provision has been made for interrupt driven buffered output to multiple devices. Extended BASIC has been constructed for use of DEC RT-11 "Load Image Files" to run under

the BASIC   system. In   addition,   there   are   FORTRAN callable
subroutines   to   provide   communication between   RT-11 FORTRAN and
extended   BASIC. They refer to   direct   access, file access,   real
time,   and system   linkage. The aim is   to operate   extended BASIC
within the DEC RT-11 system.

REFERENCE

1. Applied   Research   Laboratories   ,   Sunland   Cal.,   USA: IMMA
    Automation System.

42. A NEW TOPOLOGICAL INDEX FOR TREE GRAPHS

M. Razinger
B. Kidric   Institute of   Chemistry,   POB   380,   61001 Ljubljana,
YUGOSLAVIA

Chemical   structural   formulas   can be   represented by   graphs
(ref. 1).   The   use   of   graph   theory   in the   representation   of
molecular   structures   resulted   in   the   definition   of   numerous
topological   descriptors,   extensively   used   in   the   computer
manipulation   of   chemical   entities   (ref.   2). Some   of   them
characterize   uniquely   given   structures   and   are   exploited   in
chemical documentation   while   others, namely topological indices,
which   are not   necessarily unique for a particular structure, are
used   mainly   for   the   purpose of   various   structure - property
correlations (ref. 2). However, great discriminating power between
different   structures is a   welcome   property   also in topological
indices.
A   new   index   of   great   selectivity   is   defined   for   the
description of tree graphs. Each point of a graph is described by
three topological characteristics: its   excentricity,   weight   and
degree (ref. 1), which are multipled by each other and summed over
all points in graph. The   new index SP is   compared to three older
indices, namely   W (ref. 3), ^{1}X   (ref.   4), and Z (ref.   5).   Its
superior   discriminating   power   in   alkane   isomer   series   was
established; the example of   the 35 nonane isomers illustrates the
comparison graphically.

REFERENCES

1 F. Harary: Graph Theory, Addison-Wesley, Reading, Mass., 1972.
2 A.T. Balaban, A. Chiriac, I. Motoc, Z. Simon: Lecture Notes in
  Chemistry, No. 15 Springer, Berlin, 1980
3 H. Wiener: J.Am. Chem. Soc., 69 (1947) 17
4 M. Randic: J.Am. Chem. Soc., 97 (1975) 6609
5 H. Hosoya: J. Chem. Doc., 12 (1972) 181

43. CAI System for General College Chemistry

Donald Rosenthal, Donald Woodbury & Richard Murtagh
Department of Chemistry Clarkson College of Technology Potsdam,
New York 13676

A CAI System has been developed on a one hundred terminal IBM
4341 computer. Over one hundred CAI programs are available which
are written in VS BASIC. A monitor (or supervisor) has been
written which upon sign-on locks the student into the system. The
monitor can help the student locate programs by identifying each
program with one of thirty-five different categories and with a
chapter in the course textbook. A student can access a program
simply by typing its name. Once finished with one program a
student can proceed to another program or exit from the system.

The monitor keeps records of student performance and permits
students to leave messages for the instructor or the instructor
leave messages for the student. The system has been used for
several years by students who take a two semester general
chemistry course.

44. APPLICATION OF NIH/EPA CIS CHEMLAB PROGRAMS FOR THE DESIGN OF RADIOPHARMACEUTICALS.

W.J. Rzeszotarski, R. Potenzone, Jr.[*], W.C. Eckelman, R.E. Gibson
and R.C. Reba
Radiopharmaceutical Chemistry, George Washington University
Medical Center, Washington, D.C. 20037
[*]US Environmental Protection Agency, Washington, DC

The NIH-EPA Chemical Informational System (CIS) (ref. 1) consists of a collection of disk-stored data bases and a battery of interactive computer programs. The Chemical Modeling Laboratory (CHEMLAB) program is the third generation of the CAMSEQ (ref. 2) series of molecular processing programs and represents a package of molecular structure calculation routines tied together by a centralized controller program.

We have applied CHEMLAB programs for analysis of a group of muscarinic antagonists. The following options have been exercised:

1 Linear Free Energy with suboptions: a) octanol/water partition co-efficient (log P); b) free energy of solvation in water; c) molecular volume; d) connectivity indexes; and e) polarizability.

2 Conformational Analysis

3 Surface Areas (solvated and unsolvated).

4 Dipole Moment.

5 Shape Analysis.

Conformational analysis was performed to determine the lowest free energy conformations of the investigated 13 antagonists of general formula I. The obtained conformers were in turn compared to X-ray crystallographic studies of 3 quinuclidinyl benzilate (QNB). The conformation of one of the investigated antagonists: 3-quinuclidinyl xanthene-9-carboxylate (QNX) with a rigid skeleton, served as another reference. In addition to the calculated log P values we determined the capacity factors (k') obtained by reversed phase HPLC (ref. 3).

The data obtained from CHEMLAB, association constants obtained in vitro studies using rat ventricular muscle and rabbit caudate muscarinic receptor fractions, the results of the in vivo displacement studies, and experimentally determined k' were analysed together in search of pattern recognition that is, correlations among the above parameters.

The preliminary results permit the following conclusions:

a. 3-Quinuclidinol provides the most compact, rigid aminoalcohol moiety with the greatest accessibility of the lone electron pair on the nitrogen atom and therefore any search for a better aminoalcohol moiety will be difficult.

b. The replacement of one of the phenyl rings of QNB with a cyclopentyl group leads to substantial improvement of affinity in the heart and displaceability in the brain. When the benzylic acid part is replaced by xanethene-9-carboxylic acid, a significant difference in affinity to the receptor in the two tissues is observed. In these cases the shape, volume and lipophilicity of the molecules play the determining role.

REFERENCES
1 Milne G.W.A, Fisk C.L., Heller S.R., and Potenzone R., Jr. Science 215 (1982) 371
2 Potenzone R.,Jr., Cavicchi E., Weintraub H.J.R., and Hopfinger A.J., Comput. Chem. 1, (1977) 187.
3 Unger S.H. and Chiang G.H., J. Med. Chem. 24 (1981) 262.

45. STRUCTURE ELUCIDATION SYSTEM : CHEMICS

Shin-ichi Sasaki and Hidetsugu Abe
Toyohashi University of Technology; Tempaku, Toyohashi, JAPAN

The authors have devoted to develop the computer system for structure elucidation, called CHEMICS, since 1966. In this system, the spectroscopic data (mass, infrared, and proton and C-13) of an unknown organic compound are analyzed automatically and candidate structures for the compound are generated on the basis of the information obtained. The correct answer for the compound is always a member of the group of structures generated. This is the most characteristic feature of the system, because even an operator without experience in organic chemistry can obtain the correct structural formula for an unknown organic compound by introducing spectroscopic data. This feature does, however, tend to irritate experienced chemists because of the large number of structures that can be output, many of which are irrelevant to the problem. Chemists almost always have some additional information about unknowns. This may have been obtained by ordinary chemical observation of the sample, by further examination of spectroscopic data, by determination of organic functional groups, etc. CHEMICS had been revised so that it not only accepts any size of substructure and/or a skeletal structure available for input from other structural knowledge about the sample. With the addition of this function, CHEMICS provides more accurate structure elucidation, and coincidentally the chemist can participate in the work more closely.

However, even if such substructures originated by chemists'
discretion were added to the elucidation work, CHEMICS still
provides the candidate structures in no small numbers in some
cases. For further removal of the redundancy and to give the rank
of possibility to each of the finally survived structures, the
following two functions are endowed with the system at the last
step of the computation. Function of the removal of redundancy is
the prediction of number signals in a broad band decoupled C-13
NMR from the topological representation of structural formula by
means of the method of symmetry perception. The predicted number
of signals for each candidate structure is compared with the
observed number. After structure generation, this option may be
used to prune off redundant structures from a group of candidates
in a man-machine interactive mode. Furthermore, for putting a
ranking mark on each candidate finally survived, strain energy of
cyclic structure is estimated. There are several approaches to
estimate the strain energies, i.e., quantum mechanical
calculations, various kinds of empirical force field methods and
so on. They are not always suitable for CHEMICS because of the
large amount of computation time. The authors have developed a
simplified force field method to estimate only ring strain
energies.

A full explanation of the recently advanced system is to be
presented.

The system now has been extended to enable us to treat the
compounds with C, H, O, N, S and halogens. Detail will be
presented by one of the authors (H.A.) in the poster session.

46. REACTION SIMULATION AND REACTION PATH PREDICTION

Dr. Wolfgang Schubert

Reactions are simulated in a computer using a mathematical
model of chemistry. By repeatedly simulating reactions, reaction
paths of substances can be predicted.

Reactions are modeled as redistributions of valence electrons
of a fixed set of atoms and as redistributions of atoms in three
dimensional space. The structure of the molecules involved is
contained in a set of bonds and a set of cartesian coordinates.

By operating on the set of bonds and the cartesian coordinates reactions are simulated. The set of atoms is invariant under reactions.

Operators on the set of bonds are corresponding to the elementary reactions of shifting one electron from one atom to another and forming a bond between two atoms by one electron of each. These elementary reactions are the simplest red-ox process and the homoaptic bond forming. Breaking a bond is given by the reverse homoaptic, the homolytic reaction. All reactions can be simulated by applying a sequence of operators corresponding to these elementary reactions to the set of bonds. The set of cartesian coordinates is operated upon so as to minimize the strain energy in the molecules.

Carrying out elementary reactions and generating a reaction network is accompanied by evaluating the products according to chemical criteria. These criteria change with the type of application and may contain checks of valence limits and charge distributions, tests for forbidden substructures and steric congestion. Duplicate entries or cycles in the reaction network are detected. Mesomeric and tautomeric structures have to be recognized. Important is that chemical criteria to evaluate the products and reactions are employed as general chemical principles and not facts valid only for specific reactions. Since the algorithms used to simulate reactions are not based on known reactions, the results are not biased and the system has potentially predictive power.

The system ASSOR (Advanced Simulation System of Reactions) has been programmed according to the above principles. Results obtained by ASSOR will be discussed.

47.   COMPUTER-CONTROL OF THERMAL WEIGHT-LOSS REACTIONS TO IMPROVE
      THE KINETIC MEASUREMENT PROCESS

Wendel J. Shuely
Research Division, Chemical Systems Laboratory, Aberdeen Proving
Ground, MD 21010

Paul E. Field
Chemistry Department, Virginia Polytechnic Institute & S.U.
Blacksburg, VA 24061

A new approach for the investigation of thermal reactions is
being developed to aid in the study of thermal disposal of
hazardous compounds. A promising thermal disposal method involves
the low temperature conversion of a hazardous compound to a
product that might offer the opportunity for resource recovery or
a reduced disposal problem. Consequently, computer-controlled,
closed-loop experimentation is being developed and applied to
pesticide samples in order to measure thermal kinetic parameters
for these conversions. The concept and applications are being
presented elsewhere, therefore, the emphasis here is on the method
development, especially the computer-instrument interface and the
software control algorithm.
     The computer-controlled method decreases thermal lag errors,
provides near-equilibrium thermodynamic reaction conditions, and
furnishes a complex temperature profile that controls reaction
rates to values that can be accurately measured. Thermal and mass
fluxes are smoothed and partial compensation for sample
self-heating or self-cooling is obtained. These advantages are
gained by interchanging the usual independent and dependent
variables, temperature and weight, respectively. The software
algorithm controls the thermal weight-loss reaction rate to a
preselected value by continuously varying the temperature.
     The software was written in FORTRAN and assembly language. A
specialized proportional-integral-derivative control algorithm
with rapid scan, coast, rate limits and other modes was designed;
a module of about 15 equations and 20 decision points was required
for all modes. The calculations result in a digital word
proportional to the desired temperature change. The control

variable is converted to an analog signal, voltage amplified, inverted, amperage amplified, and routed through a STDP switch, a relay switch, resistor network, and integrating operational amplifier. This thermocouple level output signal then becomes the input to a standard TRIAC furnace control circuit.

Results with several bimolecular, heterolytic, fission reactions and pesticide decomposition reactions are presented. Graphic display of kinetic data, Arrhenius plots, data aquisition and control printout, and computer simulation results are provided.

48. TOTALLY-COMPUTER-AUTOMATED PREPARATIVE LIQUID CHROMATOGRAPHY

Francois R. Sugnaux
Department of Chemistry, Stanford University, Stanford, California 94305

Microprocessor-based laboratory instruments have provided the chemists with an easier and improved control of instrumental functions, by self-testing, automatic recalibration and storage of instrument parameters. Simplification of operation has resulted for instruments from many major companies in a monolithic design which prevents any modification of the dedicated functions.

However, the full potential of microprocessor-controlled apparatus can be best exploited in the interactive use of several modular components of a system. Therefore, innovative methods require interlinked and user-programmable instruments. We have illustrated this approach in the intelligent automation of a preparative high performance liquid chromatograph (HPLC). This application requires a particularly large versatility because of the complexity of the separations and of the great number of possible choices of solvents, injectors, columns and detectors.

The automated preparative HPLC system, based on Waters (Milford, Mass.) modules comprised two Model M45 pumps, a WISP 710B auto-sampler, a #720 System Controller, a #730 Data Module and two pneumatically actuated (2 and 6 positions) valves. The 10mm i.d. columns were filled either with silica gel or RP-18 and were connected to a Waters #401 refractometer or a Waters #450 variable UV photometer. The computer-controlled collection was

performed with a Foxy (Isco, Lincoln, Nebr.) fraction collector.
The computer program AUTOPREP, using BASIC language, was written
to upgrade the standard System Controller program and control in
an unattended way both the large volume injection and the
collection of single peaks of solutes. Sample mixtures of
unlimited volume could be injected through the pump head or,
alternatively, samples of up to 2 ml could be injected by the
auto-sampler. The peaks were defined by monitoring the first
derivative of the detector's signal, a process that has many
advantages over the usual intensity threshold peak detection. A
combination of continuously adjusted peak slope thresholds and
time windows provided logic decisions during the chromatographic
runs, which permitted to achieve various collection patterns.
finally, the model collection pattern generated by the program
could be used for repetitive injections and collections.

49. MOSA: A SYSTEM FOR MOLECULAR SHAPE ANALYSIS

Isamu Suzuki and Tsuguchika Kaminuma
The Tokyo Metropolitan Institute of Medical Science, 18-22,
Honkomagome 3-Chome, Bunkyo-Ku, Tokyo, 113 JAPAN

An Interactive Molecular Shape Analysis system (MOSA) is
developed for chemical and biomedical applications in mind. The
MOSA is designed mainly for interactive molecular display and
molecular shape analysis. It generates 3-d molecular images,
displays them both in "ball and stick" and "space-filling" models,
compares a set of 3-d molecular structures, and extracts their
common features including substructure matching and clustering.
The system is implemented on the PDP 11/70 and is feasible for
molecules whose atomic numbers are up to 250. The system has been
tested by various sets of testing objects. Various illustrative
examples will be shown on the poster.

50. COMPUTER-ASSISTED PREDICTION OF HIGHLY BIOACTIVE COMPOUNDS

Chiyozo Takayama, Mototsugu Yoshida(*), Yoshikatsu Miyashita, and
Shin-Ichi Sasaki

Toyohashi University of Technology, Toyohashi, Aichi 440, JAPAN;
(*) Sumitomo Chemical Co., Ltd., Toyonaka, Osaka 565, JAPAN

The computer-assisted design of highly bioactive compounds
among congeners is considered to include two phases. The one is
the development of a QSAR equation and the other is the decision
for the direction of the structual modification based on the
correlation, such as the selection of substituents. We have
developed computer programs called PREHAC1 and PREHAC2 which aid
in the impartial selection of substituents for the synthesis of
highly bio-active derivatives based on the Hansch-Fujita analysis.
We have first input thirteen sets of physicochemical parameter
values of abut 400 useful substituents in a data file as Master
Data. For each substituent, 4 substitution positions, that is,
aliphatic, ortho, meta, and para positions, were considered.
Substituents were classified into 4 categories according to the
situations such as synthetic source, difficulty of synthesis,
frequency of occurrence, etc. Each category was indicated by an
index called Popularity Order. The PREHAC1 program can be used for
the design in the series of aliphatic and aromatic monosubstituted
derivatives. This program is activated by the input of the
coefficients and intercept of the QSAR equation with the
corresponding parameter codes, the information on the number of
the substituents to be printed out, etc. Then Master Data is
searched. Calculated activity values are ranked in descending
order and the high-ranking substituents are printed out by the
specified number. The extended program, PREHAC2, is for the series
of aromatic disubstituted derivatives. In this program, the
Popularity Order index is used effectively in order to restrict
the combinations of the two substituents. Some examples will be
presented to illustrate the PREHAC programs.

51. METHODS OF ORBIT-GRAPHS

Masahiro Uchino
Laboratory of Resources Utilisation, Tokyo Institute of Technology
4259 Nagatsudacho, Midoriku, YOKOHAMA

Algorithms for unique coding of chemical graph are required not only for chemical structure data base but also for eliminating duplicated structures in structure-generation. Many algorithms reported so far fail to give or recognize the canonical codes of large regular graphs.

To overcome this difficulty, we must solve two problems:
1) How to reduce the number of possible representations generated by the numbering rule adopted for vertex-numbering of a graph.
2) How to avoid the duplicated numberings due to symmetry of a graph

These two problems have been solved on the group theoretical basis and we propose the "Methods of Orbit-Graphs" as the method for unique coding as well as for computation of bond-atom symmetry. In the proposed method, two new conceptions "fitness of node for the number i" and "orbit-graph of a chemical graph" are introduced in order to reduce the number of numberings to be examined. The "orbit-graph" is a directed graph which can be easily derived from chemical graph to be coded (i.e., the number of permutations required for determination of "orbit-graph" is less than 1 + sum of out-degrees of nodes in the "orbit-graph") and contains all information about symmetry which is necessary and sufficient for coding of a chemical graph. This directed graph is used in combination of the conception "fitness" to provide a small number of numberings which contain a canonical numbering.

The usefulness of the proposed method is essentially independent of the definition of the canonical code and if we employ the"orbit-graph", Morgan's name of the dodecahedron can be easily determined.

52. AUTOMATED SYSTEM FOR ENCODING AND INPUT OF CHEMICAL STRUCTURE INFORMATION TO A COMPUTER

G. P. Ulyanov
Scientific Information Centre for Molecular Spectroscopy,
Institute of Organic chemistry, Siberian Branch of the USSR
Academy of Sciences. Academician Lavrentiev Avenue, 9, 630090,
Novosibirsk (USSR)

Main features and capabilities of multiterminal chemical
structure graphical input system designed for the creation of
large data bases are discussed. The hardware of the system
consists of specialized graphical encoding tablets (the so-called
"GRAF"), minicomputer and some CAMAC devices.

Graphical encoding tablet is used for input of chemical
structures by means of drawing. After drawing, a structural
formula is transformed to a connectivity matrix of chemical graph
and a list of two-dimensional coordinates of atom positions in the
picture. The system provides simultaneous input of graphical and
alphanumeric information from 8 tablets, real time checking of
input data and operative visualization of the input structures in
color graphic CAMAC-equipped displays.

There are the following options in the system: input,
structure editing, and recording of correct structure codes on
magnetic tape. The average output of an encoding tablet is about
20 structures per hour. Minicomputer used in the system is
compatible with the pDp-11 series.

53. EDUCATION OF DATA PROCESSISNG IN ANALYTICAL CHEMISTRY

Gabor E. Veress, Erno Pungor
Technical University of Budapest, Budapest, HUNGARY

The logical construction of the education of data processing
is based on the application of system theory in analytical
chemistry. From the point of view of system theory analytical
chemical methods can be regarded as systems producing chemical
information. From the point of view of semiotics production of
chemical information is based on a semiosis, i. e. analytical

chemical   systems   consist   of signal production and   signal
interpretation.

Analytical chemical signal interpretation consists of
analytical signal processing and analytical correlation. The
function of signal processing, is the production of analytical
information from the analytical signal. The function of analytical
correlation means on one hand the construction of the correlation
between analytical and chemical information /calibration/ and on
the other hand the use of the correlation, i. e. analytical
inference.

Based on these system theoretical backgrounds the education of
data processing means the education of different signal processing
methods /signal analysis and processing, transformations, enroe
fitting, feature selections, etc./ and of different analytical
correlation methods /multicomponent analysis, standard addition,
rank   annihilation,   Kalman-filter,   pattern   recognition
classification, data banks, etc./ and of the characterization of
analytical chemical systems /statistical, information theoretical,
decision theoretical, etc./ In practice this education system
seems very effective and useful.

54. APPLICATION OF A GLOBAL OPTIMIZATION ALGORITHM TO DATA PROCESSING IN CHEMICAL EXPERIMENTS

Q. Wang, C. Feng and Y. Hui
Shanghai Institute of Organic Chemistry, Academia Sinica, CHINA

The experimental data fitting can be made by using parameter
estimation method. The measure of fitting is usually the least
square criterion (the sum of square of difference between the
experimental and calculated values). Mathematically, it reduces to
an optimization problem. Conventional optimization algorithms
(local algorithms) require proper estimation of initial values,
incorrect initial values often lead to unsatisfactory or even
divergent results. Moreever, they require more core memories to do
such computation and the extreme obtained are often local extreme
instead of global ones. A new global optimization algorithm(1)
reported here can overcome such disadvantages to a large extent.

Let F(x) be a continuous function in the region G of n-dimensional Euclidean space R^n

$$G = \{ x | a_i \leq x_i \leq b_i , \quad i = 1, 2, \ldots n \}$$

The algorithm for calculating the global minimum and the global minimum point of F(x) in G is as follows:

1) Given the number of initial sampling point P, the number of the reserved points S and the searching parameter E.
2) in the initial searching range (n-dimensional cuboid D), P points of x are created randomly.

$$x_i = a_i + (b_i - a_i) \cdot R_i \quad i = 1, 2, \ldots, n$$

Where R_i is pseudo-random number distributed uniformly in (0,1). Calculating F(x) at every point, reserving S points with the least function values out of P points. Storing them in ascending order

$$F_1 \leq F_2 \leq \ldots \leq F_s \quad \text{and calculating their average } \bar{F}:$$

$$\bar{F} = \frac{1}{S} \sum_{k=1}^{S} F_k$$

3) Construction the least n-dimensional cuboid D_1 which just covers S reserved points. If the point with the least junction value is situated at the center of D_1 , adjusting the boundaries of D_1 toward the direction of searching for the minimum.

4) Repeating step 2) with D_1 being the searching range, updating S reserved points until

$$F_s < \bar{F}$$

going to step 3) to obtain a new searching

$$D_2 \ldots$$

the process is repeated and the searching range is getting smaller and smaller.

$$D \supset D_1 \supset D_2 \supset D_3 \ \cdot \ \cdot \ \cdot \ \supset D_k$$

5) When the volume (or edge length) of D_k is less than E, the computation is ended. The extreme point and extreme value x and F(x) are found simultaneously. This algorithm is free of divergent problem, and does not need accurate initial values. It requires few core memories and so can be implemented on microcomputers. It has been used successfully to resolve overlapping peaks in spectra and evaluate rate constants of complicated chemical reactions in our laboratory.

55. APLICATIONS OF AN INTERACTIVE VECTOR GRAPHICS SYSTEM TO CHEMISTRY

J. Weber, J.J. Combremont and M. Roch
Laboratory of Computatioal Chemistry, University of Geneva, 30 quai Ansermet, 1211 Geneva 4, SWITZERLAND

For four years now, we have been using an interactive computer graphics equipment dedicated to chemical applications. Our configuration consists mainly of a Vector General 3D calligraphic (black and white) system linked to a PDP 11/60 host computer. The applications we have developed belong to both fields of chemical research and education. Indeed, several important concepts of chemistry require molecular modeling or visualization of reaction mechanisms and computer graphics is an ideal technique to this end.

    This poster, together with a 16mm film which will be presented during the movie session, illustrates the following applications: (i) interactive visualization of molecular properties derived from quantum chemical calculations of Huckel type; (ii) representation of an important class of dynamic processes in molecules called molecular rearrangements. The latter application is complicated by the fact that a geometrical data base is needed for representing the chemical reaction path and this requires generally to perform sophisticated quantum chemical calculations. The film presents three molecular rearrangements: reversible formation of cyclobutene from 1,3-butadiene, intramolecular migration of a methyl group over an allyl substrate, pseudo-rotation of a seven-membered heterocyclic compound. In each case, in addition to the 3D molecular reaction paths, the stereochemistry and/or energy implications of these processes are dynamically represented. The simultaneous visualization of these concepts should help the chemist evaluating the chemical reactivity of these compounds in terms of their possible molecular structures.

56. PREDICTION OF STRUCTURE/SELECTIVITY RELATIONSHIPS OF ION CARRIERS

Martin Welti and Erno Pretsch
Department of Organic Chemistry Swiss Federal Institute of Technology CH-8092 Zurich, SWITZERLAND

Using a previously described additive procedure (refs. 1-3) the results of abinitio computations on cation-ligand interactions can be approximated. The interaction energy between a ligand molecule and an ion is thereby calculated as a sum of atom-ion pair interactions. The contribution of each atom of the ligand molecule is described by a three parameter function consisting of a Lennard-Jones potential and a Coulombic term. Corresponding parameters for the interaction of lithium-, sodium-, magnesium-, and calcium- ions with carrier molecules were derived on the basis of ab initio calculations on complexes of small ligand molecules for H, C, N, O, and S atoms in different chemical environments refs. 2-5). Here we present results of the application of this model on a series of model complexes with known structures.

REFERENCES

1 E. Clementi, Computational Aspects for Large Chemical Systems, Lecture Notes in Chemistry, Nr.19 Springer-Verlag, Berlin 1980.
2 G. Corongiu, E. Clementi, E. Pretsch and W. Simon, J. Chem. Phys. 70, (1979) 1266.
3 G. Corongiu, E. Clementi, E. Pretsch and W. Simon, J.Chem.Phys 72, (1980) 3096.
4 E. Pretsch, J. Bendl, P. Portmann and M. Welti, in "Steric Effects in Biomolecules", G. Naray-Szabo (Ed.), Elsevier - Akademiai Kiado, in press.
5 M. Welti, E. Pretsch, E. Clementi and W. Simon, in preparation.

57. TEACHING COMPUTERIZED CHEMICAL INFORMATION

Gary Wiggins

There is a need to teach computer-based chemical information techniques to students and faculty in academic institutions. A survey of the printed workbooks and online learning aids available to accomplish this task is presented. One department's experience in this area is described, as is a degree program leading to the Master of Library Science-Information Specialist in Chemistry degree. Some problems in implementing courses on the subject in academic institutions, including lack of equipment, lack of suitable textbooks covering the full range of services currently available, and lack of cooperation from database producers and vendors, are also covered.

58. EXPLICIT LOGICAL REASONING IN SYNTHESIS

W. T. Wipke and D. P. Dolata
Department of Chemistry, University of California, Santa Cruz, CA 95064

Designing an organic synthesis is an intellectual enterprise requiring vast bodies of knowledge, careful analysis, and clear reasoning. To provide the ability for a machine to perform such reasoning one must provide the knowledge and an inference engine to use it. We have created an inference engine (QED) based upon the First Order Predicate Calculus, and are writing some of the basic concepts of synthesis design into axioms in this calculus language. The application of these axioms to specific molecules creates chains of inference leading to a plan based upon careful analysis for the molecules specific nature, rather than upon an ad-hoc generalized procedure created at some earlier date. Thus the plans are each 'tailored' for the molecule in question.
The explicit representation of axioms using common chemical terminology and logical connectives makes the knowledge base easy to understand and to extend. We are using QED as an exploratory tool to discover these 'synthetic' axioms. We started with a simple knowledge base of obvious axioms. By allowing the inference

engine to make suggestions for specific molecules, we discover where the knowledge is incomplete or inconsistent. Since we know that the inferences generated by the First Order Predicate Calculus must be sound, any incompleteness or inconsistencies between the observed and expected results must be the result of the knowledge base. By successive refinements of the knowledge base, we will be able to create a 'spanning set' of axioms which covers accepted practice, and also points the way to new methods and schemes.

59. ISOTOPE CATALOGS ONLINE

William J. Wiswesser
USDA Weed Science Research Labs., Frederick, MD 21701

The availability of full typewriter keyboards at online terminals has stimulated a fruitful ripening of the Wiswesser Notation (WN). This standardization of Josef Loschmidt's 1861 line-formula notations was introduced internationally by an invited paper in CHEMICAL AND ENGINEERING NEWS (August 1952), and enjoyed improvements from some 700 students, teachers and practicing chemists in the 1961-64 Decoding Evaluation supported by the National Science Foundation. Research with WYLBUR files of 6600 up to 30,000 records (in 5 ring-sorted subfiles) in 1981 led to self-evident isotope suffixes that are (1) immediately recognized by the unique SPACED SHARPMARK " #"; and searchable by (2) the atomic symbol or (3) mass number in many simple compounds; or by (4) substructural subgroups or (5) chain and ring positions. The latter features are payoffs with the distinctive lower-case letter set and the substructural symbols in the advanced Wiswesser Notation (AWN), which still shares thousands of structure identities with the internationaly controlled WLN. Reprints from current reports explain the "A,B,C's" of the notation, and online searches through large WYLBUR files will demonstrate the very high high substructural discriminating power of the full online keyboard.

60. ANALYZING ANTIBIOTICS BY PAIRS – A COMPUTERIZED INFRARED SPECTRAL INTERPRETER

Hugh B. Woodruff
Merck Sharp & Dohme Research Laboratories, P.O. Box 2000
Rahway, NJ 07065

Infrared spectroscopy provides a unique fingerprint of compounds and has proven to be extremely valuable in the study of antibiotics. To interpret an infrared spectrum effectively, one usually employs an aid such as a correlation chart. Even the most comprehensive correlation charts are inadequate to evaluate completely the spect of complex structures like antibiotics which have a great diversity of functional groups. A variety of computerized techniques have been implemented to assist the interpretation of spectra. One especially promising technique is PAIRS (Program for the Analysis of IR Spectra).

This program was initially described two years ago at the Fifth International Conference on Computers in Chemical Research and Education in Toyohashi, Japan. PAIRS attemps to parallel as closely as possible the reasoning a spectroscopist uses in interpreting infrared spectra. With this statement of the underlying principle of PAIRS, it is evident that the approach is very different from conventional pattern recognition techniques. Pattern recognition treats data as an abstract vector, openly ignoring chemical intuition or prejudices. PAIRS attempts to exploit scientific principles and intuitions as much as possible. PAIRS is a heuristic approach that attempts to mimic techniques employed by humans for interpreting spectra. The precise definition of heuristic from Webster's Third New International Dictionary is "serving to guide, discover or reveal; valuable for stimulating or conducting empirical research but unproved or incapable of proof." PAIRS does indeed attempt to "reveal" possibilities from an interpretation that a scientist might have overlooked; it attempts to "stimulate" the scientist's thinking.

Since the initial work reported in Toyohashi, PAIRS interpretation rules have been tested and upgraded extensively. In a recent study, nearly 200 representative antiotic spectra were digitized and interpreted by PAIRS with very encouraging results. The presence or absence of 56 functionalities were predicted by

PAIRS for each spectrum. The correct designation was made by PAIRS 88% of the time. Details of this study will be described in the the presentation.

61. INCORPORATING THE NIH/EPA MSSS INTO THE UNDERGRADUATE CHEMISTRY CURRICULUM

Y. Wolman
Department of Organic Chemistry, The Hebrew University of Jerusalem, Jerusalem, ISRAEL

Computer usage and application in chemical curriculum have been increased dramatically during the last few years. We wish to report about our experience in introducing the chemistry major students to computerized data banks, their structure and usage. This introduction was carried out within the framework of our Advanced Organic Chemistry Lab.

The Advanced Organic Chemistry Lab is an elective lab which is given to our chemistry students in their 3rd year of studies. The aim of the lab is to teach and introduce the students to the latest developments in organic chemistry techniques and their usage as research tools.

We have decided to incorporate into our lab curriculum the usage and structure of the NIH/EPA MSSS within the framework of our mass spectrometry experiment.

The aim of the experiment is to demonstrate the usage of the mass spectrometer as an analytical as well as an experimental tool. The students learned the various features of the instrument (Varian MAT-311), ran a spectrum of an unknown compound, learned about numerical data (problems concerning retrieval of evaluated numeric data, data banks, their structure and usage with particular emphasis on the NIH/EPA MSSS). Following the lab period each student has to submit a paper on a selected topic related to the mass spectrometer as a research tool, elucidate the structure of his unknown both manually and with the aide of the MSSS and compare both results.

Using the MSSS enables the student to use the computer in an interactive way for solving lab problems. The students are getting to know some of the databank features - e.g. searching for peaks, searching for peaks and molecular weight, entering a spectrum, searching with a complete spectra. The students are divided into small groups (5-6 students per group). Each group meets together and each student spends about 8-10 min. using the terminal, followed by a group discussion. Results were compared during the discussion, search techniques were examined and a complete spectra search was run.

The students response was very enthusiastic, one could feel their interest by their reactions during and after the lab session. The group meetings went much longer than scheduled due to the many questions, queries and long discussion that followed the online searching.

62. A IR SPECTRUM DATA BASE WITH PEAK INTENSITY INFORMATION AND ITS SEARCHING STRATEGIES

C. Zheng, Y. Wand, C. Qian, C. Nie, And Y. Hui
Shanghai Institute of Organic Chemistry, Academia Sinica, CHINA

In the Shanghai Institute of Organic Chemistry, Academia Sinica, 34,000 standard IR grating spectra and 10,000 commercial IR spectra from different sources have been coded to establish a computer readable data base which contains peak position as well as intensity, half height width and other related informations. A library of 10,000 spectra from them can be searched by means of a flexible search system, named OCIRS (Organic Compound Infrared Spectrum Search System), on a low-range mini-computer. Systematic search tests give valuable estimations of various search strategies.

In the library, peak position values are taken in the range 400 to 4,000 cm^{-1}, and are 22 bytes bitscreen. The intensity and the half height width are digitized to 4 bit and 2 bit data respectively. Literal information consists of compound names (62 bytes), sources of spectra (4 bytes) and methods of sample preparation (2 bytes). The molecular formulas and molecular weights are also stored and are used as search keys. By using spectrum information only or in combination with MF and MW, the OCIRS can searh the library through several strategies in response to different user requirements.

Excluding artificial mistakes, the eventual data errors in a computer readable condensed IR data base can be expressed as E=E1+E2+E3, where E1 is the contribution from sample impurity, E2 is the deviation caused by different instruments and experimental conditions and E3 is the round-off error.

From the point of view of a search system, the error E is equivalent to the external noise of the system. Non-zero E may disturb the search result by missing, by incorrect hit or by proposing an unsuitable order of answers.

The search methods adapted by the OCIRS are: error tolerated intensive peak position index search (A), peak position match (B), peak position and intensity match (C), and the combined methods of (A), (B), (C), which lead to high accuracy and well ordered answers while the response time remains in an acceptable real-time range.

In method (A), the peaks with its intensity $T>T_{max}+3$ of every library spectrum are "vibrated" (3) and then indexed according to their positions. The index file is searched by the positions of search spectrum's intensive peaks with the intensities $T>T_{max}+1$, where T_{max} is the most intensive peak of the same spectrum. Therefore, this method is capable of tolerating deviations of peak position and intensity to avoid missing caused by E in index search.

In method (B), peak positions are the only factor to be considered in the calculation of matching coefficient (ref. 1), which follows the formula suggested by ZUPAN (ref. 3).

In method (C), the peaks of a search spectrum are compared one by one to that of library spectra and the matching coefficients are calculated from peak positions and intensities according to the formula suggested by the authors (ref. 2). The library spectra with M>140 are eliminated from the answer list.

At least 3 factors should be taken into account for estimating a spectrum search strategy: 1. Accuracy $A=\dfrac{Nf}{Ns}$, in which N_s is the total times of searching a spectrum available in the library, N_s is the times searchs which give the correct answer, even though it is not listed at the first place of the answer list. 2. Response time $T_r=t_i$, here t_i is the response time of i-th search. 3. Signal/noise ratio of the search system $R=\dfrac{\Sigma M_i'}{\Sigma M_i}$, where M_i is the matching coefficient of the correct answer M_i' is the minimum one of other answers in the answer list of the search and

Σ Mi represents the level of noise transferred by the search system.

Two types of search tests have been made for different search methods (ref. 2).

For 1236 library spectra of fluorides, tests of type 1 (85 searchs) and type 2 (5 searchs) of method B are made (ref. 1).

For about 8,000 library spectra, other 2 strategies which use combined search methods (A+B) (A followed by B) of 40 searchs and (A+C) method of 60 searchs are tested. In both cases, A=1. For (A+C), the correct answers are always listed in the first place of the answer lists.For (A+B), here are 2 exceptions.

The fact that evidently unsuitable answer orders are sometimes given by (A+B) method can be rationalized by the lower S/N ratio of (A+B) method which has a value 2.6, as compared to R=3.5 value of the (A+C) method.

The tested performances of B, (A+B) and (A+C) are collected in Table 1, which reveals the advantage of (A+C) over (A+B) in proposing a properly ordered answer list and the outstanding searching efficiency of (A+B) method.

Table 1

Tested item	Method		
	(B)	(A+B)	(A+C)
Accuracy A	+	+	+
Signal/noise ratio R	-	-	+
Response time T_r	vp	+	-

"+": good;   "-": poor;   "vp": very poor

REFERENCES

1 Zheng Chong-zhi, Wang Yuan, Qian Cheng, Nie Chong-shi & Hui Yong-zheng Kexu Tongbao, 27 (1982) 396.

2 Zheng Chong-zhi et al., "OCIRS infrared spect. searching system" (unpublished)

3 Zupan, J., et al., Computers and Chemistry 1 (1977) 77.

63. LARGE HIERARCHICAL TREE FOR A STRUCTURE ELUCIDATION SYSTEM

J. Zupan and M. E. Munk
Chemistry Department, Arizona State Universiy, Tempe 85287

The poster presents preliminary results on the use of hierarchical trees of infrared spectra for the prediction of structural fragments. The part of the INTERPRET program of system CASE (ref. 1) that handles the infrared spectra is described and discussed. The method employed is based on the recently developed procedure for generation of hierarchical trees.(ref. 2) The data base used in the study consists of 500 infrared spectra fully digitized (651 points per spectrum) and after transformed into a 100 dimensional, complex Fourier space. The hierarchical tree used for the selection of substructures was generated with 300 carefully selected infrared spectra, while the other 200 spectra were used for testing of the prediction ability of the tree. The generated hierarchical tree is described in detail. For the concise and fast description of compounds structures the Wiswesser Line-Chemical Notation was used.

REFERENCES

1 C.A. Shelley, M.E. Munk, Anal. Chem. Acta, 133 (1981) 507-516
2 J. Zupan, Clustering of Large Data Sets, Research Studies Press
   Press (John Wiley), Rochester, 1982

List of Attendees

Abe, Hidetsugu, Research Center for Chemometrics, Toyohashi University of Technology, Hibarigaoka, Tempaku, Toyohashi 440, JAPAN

Abrahamson, Earl, E.I. Dupont De Nemours, Central Research and Development, Wilmington, DE 19898

Abrahamsson, Sixten, Dept. of Structural Chemistry, University of Goteborg, S-40033 Goteborg, SWEDEN

Abu-Dari, Kamal, Chemistry Department, University of Jordan, Amman, JORDAN

Albright, Mike, JEOL, 235 Birchwood Avenue, Cranford, NJ 07016

Allen, Michael J., Glaxo Group Research Ltd., Greenford Road, Greenford, Middlesex, ENGLAND

Attias, Roger, Lab de Chemie Org Phy, Universite Paris VII, 1 Rue Guy de la Brosse, 70005 Paris, FRANCE

Billet, Lucien, Rhone-Poulenc Recherches, Centre de Recherches de Saint-Fons, 85, Avenue des Freres Perret, Boite Postale 62, 69190 - Saint-Fons, FRANCE

Blower, Paul E., Chemical Abstracts Service, P. O. Box 3012, Columbus, OH 43210

Bober, Al, Fein-Marquart Associates, 7215 York Road, Baltimore, MD 21212

Boyle, Harry F., Online Services, Chemical Abstracts Service, P.O. Box 3012, Columbus, OH 43210

Bremser, Wolfgang, BASF Aktiengesellschaft, D-6700 Ludwigshafen/Rhein, GERMANY

Bright, David S., Chemistry B326, National Bureau of Standards, Washington, DC 20234

Bruck, Peter, II Pusztaszeri ut 57/69, MTA KKKl Budapest, HUNGARY

Brunner, Thomas R., Wyeth Labs, P.O. Box 8299, Philadelphia, PA 19101

Buttrill, W. Hearon, Fein-Marquart Associates, 7215 York Road, Baltimore, MD 21212

Carhart, Ray, Lederle Laboratories, 500 N. Middletown Road, Pearl River, NY 10965

Carter, Forrest L., Naval Research Laboratory , Chemistry Division, Bldg. 207, Code 6171, Washington, DC 20375

Choplin, Francois, Rhone-Poulenc Recherches, Centre de Recherches
    de Saint-Fons, 85, Avenue des Freres Perret, Boite Postale
    62, 69190 Saint Fons, FRANCE

Citroen, Charles L., Centre for Information - TNO, P.O. Box 36,
    2600 AA Delft , THE NETHERLANDS

Clerc, J. T., Universitat Bern, Pharmazeutishes Institut, CH-3012
    BERN, Sahlstrasse 10, SWITZERLAND

Collins, Ronald W., Eastern Michigan University, Pierce Hall 138,
    Ypsilanti, MI 48197

Culp, F. Bartow, 3M Corporation, Technical Planning &
    Coordination, 201-2C-12 3M Center, St. Paul, MN 55144

De Clercq, Pierre J., State University of Ghent, Laboratory for
    Organic Chemistry, Krygslaan 271 (S4) B-9000 Gent, BELGIUM

de Maine, P.A.D., Electrical Engineering Department, Dunstan Hall,
    Auburn University, AL 35899

Delaney, Mike, Boston University, Department of Chemistry

Derendiaev, Boris, Inst. Org. Chem., Siberian Div. of Academy
    Sciences, 90 Novosibirsk 630090, USSR

Dessy, Raymond E., Virginia Polytechnic Institute & , State
    University, Chemistry Department, Blacksburg, VA 24061

Devon, Trevor K., Pfizer Central Research, Sandwich, CT13, 9NJ,
    Kent, ENGLAND

Dimitrov, Valentin, Institute of Organic Chemistry, Bulgarian
    Academy of Sciences, 1113, Sofia, BULGARIA

Dittmar, Bruce I., DuPont Company, Experimental Station,
    Biochemicals Department, Bldg. 324, Room 105, Wilmington, DE
    19899

Dobosh, Paul A., Chemistry Department, Mount Holyoke Campus, South
    Hadley, MA 01075

Dominy, Beryl W., Pfizer Inc., Eastern Point Road, Groton, CT
    06340

Domokos, L., Technical University, H-1111 Budapest, Gellert ter 4,
    HUNGARY

Dreiding, Andreas, Organisch-Chemisches Institute, Universitat
    Zurich-Irchel, Winterthurerstrasse 190, CH-8057 Zurich,
    SWITZERLAND

Dubois, Jacques E., Lab de Chemie Org Phy, Universite Paris VII, 1
    Rue Guy de la Brosse, 70005 Paris, FRANCE

Dumoulin, Charles, Syracuse University, Department of Chemistry,
    Bowne Hall , Syracuse, NY 13210

Eakin, Diane R., CRC Systems Inc., 4020 Williamsburg Court, Fairfax, VA 22032

Elder, Michael, Daresbury Laboratory, SERC, Daresbury, Warrington WA4 4AD, UNITED KINGDOM

Enslein, Kurt, Health Designs, Inc., 183 Main Street E, Rochester, NY 14604

Farkas, Margit, Institute Neviki, H-8201 Veszprem, P. O. Box 39, HUNGARY

Fassler, Dieter, Friedrich-Schiller Universitaet, Jena-Sektion Chemie, GDR-6900 Jena, A.M. Steiger 3, German Democratic Republic

Feldman, Al, National Cancer Institute, Blair Bldg., 416, 8300 Colesville Road, Bethesda, MD 20205

Feldmann, Richard J., DCRT, NIH, Bldg. 12A, Room 3001, Bethesda, MD 20205

Feuer, Bernice, Celanese Research Company, 86 Morris Avenue, Summit, NJ 07901

Figueras, John, Kodak Research Laboratories, Building 82, Rochester, NY 14650

Fincham, David, DAP Support Unit, Computer Centre, Queen Mary College, Mile End Road, London E1 4NS, UNITED KINGDOM

Fisanick, William, Chemical Abstracts Service, 2540 Olentangy River Road, P. O. Box 3012, Columbus, OH 43210

Floersheim, P., Organisch-Chemisches Institute, Universitat Zurich-Irchel, Winterthurerstrasse 190, Ch-8057, Zurich, SWITZERLAND

Frank, Ildiko, University of Washington, Department of Chemistry, BG-10, Seattle, WA 98195

Frei, K., Preclinical Research, SANDOZ, Basel, SWITZERLAND

Fuller, Allan, Nicolet Analytical Instruments, 5225-1 Verona Road, Madison, WI 53711

Gasteiger, Johann, Institute of Oranic Chemistry, TU Munich, D-8046 Gerching, WEST GERMANY

Gay, J. P., DARC-Paris, 193, rue de Bercy, 75012 Paris, FRANCE

Gelernter, Herb, Department of Computer Science, SUNY - Stony Brook, Stony Brook, NY 11794

Gift, Jeff, Hazardous Chemicals Staff, US Coast Guard, Pollution Response Branch, 2100 2nd Street, SW, Washington, DC 20593

Gilbert, Richard, University of South Florida, Chemical and Mechanical Department, Tampa, FL 33620

Goddard, John D., Scientific Numeric Databases, Canada Institute
    for Scientific &, Technical Information, Montreal Road, Bldg.
    M-55, Ottawa, Ontario, KIA OS2, CANADA

Gotkis, Judith K., Wyeth Laboratories, Inc., P.O. Box 8299,
    Philadelphia, PA 19101

Grethe, Guenter, Hoffman-La Roche Inc., Nutley, NJ 07110

Griffin, Stephen M., National Science Foundation, 1800 G Street
    NW, Washington, DC 20550

Grotch, Stanley L., Lawrence Livermore National, Laboratory, P.O.
    Box 808, L-329, Livermore, CA 94550

Hammer, Charles F., Chemistry Department, Georgetown University,
    Washington, DC 20057

Haraki, Kevin S., Lederle Laboratories, American Cyanamid Co.,
    Pearl River, NY 10965

Harmon, Robert E., Department of Chemistry, Western Michigan
    University, Kalamazoo, MI 49008

Harry, John B., Chemical Abstracts Service, 2540 Olentangy River
    Road, P. O. Box 3012, Columbus, OH 43210

Hasson, Marsha, CRC Systems, Inc., 4020 Williamsburg Court,
    Fairfax, VA 22032

Hazard, George F. Jr., National Library of Medicine, Specialized
    Informations Services, BISB, 8600 Rockville Pike, Bethesda,
    MD 20209

He, Pu, Institute of Chemistry, Academia Sinica, Peking, CHINA

Heller, Stephen R., US Environmental Protection Agency, MIDSD,
    PM-218, 401 M Street, SW, Washington, DC 20460

High, Joe B., Chemical Information Specialist, Stratford, Apt.
    1319, 4901 Seminary Road, Alexandria, VA 22311

Hilderbrandt, National Science Foundation, Program Officer for
    Structural,   Chemistry and Thermodynamics, Division of
    Chemistry, Room 340, Washington, DC 20550

Hilsenrath, Joseph, Office of Standard Reference Data, National
    Bureau of Standards, Room B327, Bldg. 221, Washington, DC
    20234

Hippe, Zdizslaw, Technical University, Priv. Ossolinskich St., 5,
    P. O. Box 85, Rzeszow 35-959, POLAND

Hooley, David J., The Standard Oil Company, 4440 Warrensville
    Center Rd., Cleveland, OH 44128

Howery, Darryl G., Department of Chemistry, Brooklyn College of
    CUNY, Brooklyn, NY 11210

Hui, Yongzheng, Shanghai Institute of Organic, Chemistry, Academia Sinica, 345 Lingling Lu, Shanghai, CHINA

Hyde, Ernie, Fraser Williams Ltd., Scientific Systems, Glendower House, London Road South, Poynton, Cheshire, ENGLAND

Iizuka, Takeshi, Department of Chemistry, Faculty of Education, Gunma University, 1375 Aramaki-machi, Maebashi-shi, Gunma-ken, JAPAN

Ishida, Yoshiaki, UBE Industries, Ltd., Analytical Research Department, Central Research Lab, Kosushi, Ube, Yamaguchi, JAPAN

Iwasawa, Kazuo, Kinokuniya Co, Ltd., Village 101 Bldg., 1-? Sakuragaoka-Machi, Shibuya-Ku, Tokyo, JAPAN

Johnson, Jodie V., University of Florida, 314 Leigh Hall, Gainesville, FL 32611

Johnson, David, US Environmental Protection Agency, OPTS/OTS (TS-796), Washington, DC 20460

Jurs, Peter, Department of Chemistry, 152 Davey Laboratory, Penn State University, University Park, PA 16802

Kao, James, Philip Morris Research Center, P.O. Box 26583, Richmond, VA 23261

Kaufman, Joyce J., Department of Chemistry, John Hopkins University, Baltimore, MD 21218

Kennedy, John W., Research Institute, Advanced Medical Products, The Master's Lodge, Flemish Cottages, Dedham, Essex, ENGLAND

Kern, C. William, National Science Foundation, 1800 G Street, NW, Washington, DC 20550

Klopman, Gilles, Department of Chemistry, Case Western Reserve University, Cleveland, OH 44106

Koch, Kay F., Eli Lilly & Company, Lilly Research Laboratories, 307 E. McCarty Street, Indianapolis, IN 46285

Kohler, Irmgard, German Cancer Research Center, Im Neuenheimer Feld 280, 6900 Heildelberg, GERMANY

Koptyug, V., Institute for Organic Chemistry, pr. Nauki 9, Siberian Division of the, Academy of Sciences, 90 Novosibirsk, 630090, USSR

Kornbluth, Michael M., Georgetown University, Chemistry Department, 37th & "O" NW, Washington, DC 20057

Koschmann, Timothy, Abbott Laboratories, D-409, AP9-1, 14th & Sheridan Road, North Chicago, IL 60064

Kurkela, Kauko, MEDIPOLAR/Farmos Group Ltd., SF-90650 Oulu 65,
     FINLAND

Laitinen, Sauli, Technical Research Centre of Finland, Technical
     Information Service, Vuorimiehentie 5, SF-02150 ESPOO 15,
     FINLAND

Legrand, M., Roussel UCLAF Research Centre, 93230 Romainville,
     FRANCE

Liang, Xi-Yun, Institute of Chemistry, Academia Sinica, Peking,
     CHINA

Lipkowitz, Kenneth B., Department of Chemistry, Indiana-Purdue
     University, 1201 E. 38 Street, Indianapolis, IN 46205

Lipnick, Robert L., US Environmental Protection Agency, OTS/HERD
     (TS-796), 401 M Street SW, Washington, DC 20460

Lobbestael, Sandie, Warner Lambert, 2800 Plymouth Road, Ann Arbor,
     MI 48106

Long, Alan K., Harvard University Chemistry Dept., 12 Oxford
     Street, Cambridge, MA 02138

Lowry, Stephen R., Nicolet Instrument Co., 5225-1 Verona Road,
     Madison, WI 53711

Lykos, Peter, Department of Chemistry, Illinois Institute of
     Technology, Chicago, IL 60616

Malinowski, Edmund R., Dept. of Chem. & Chem. Eng., Stevens
     Institute of Technology, Hoboken, NJ 07030

Marshall, Richard A., Goodyear Tire & Rubber Company, 142 Goodyear
     Blvd., Akron, OH 44316

Marsili, Mario, Computer Chemistry, Area della Ricera di Rome CNR,
     00016 Monterotondo Scalo Rome, ITALY

Martin, Yvonne C., Abbott Laboratories, Medicinal Chemistry
     Department Rl, North Chicago, IL 60064

Martinsen, Dave, Fein-Marquart Associates, 7215 York Road,
     Baltimore, MD 21212

Matta, Lygia B., American University, Chem. Info. Systems, 5511
     Westbard Avenue, Bethesda, MD 20816

Meyer, Ingrid B., EPA/OTS/MSD/SDB, TS-793, 401 M Street SW,
     Washington, DC 20460

Mikhailova, I., Institute of Organic Chem., Academy of Sciences,
     90 Novosibirsk 630090, USSR

Miller, James A., Fein-Marquart Associates, 7215 York Road,
     Baltimore, MD 21212

Mills, O. S., Department of Chemistry, University of Manchester, Manchester M13 9PL, UNITED KINGDOM

Mills, Johnathan, SERC Rutherford Appleton Lab, Chilton, Oxon OX11 OQX, UNITED KINGDOM

Milne, G. W. A., National Cancer Institute, Blair Bldg., 416, 8300 Colesville Road, Bethesda, MD 20205

Miyagawa, Seinosuke, Kwan sei Gakvin Univ. School of Science, Nishino miya Hyogo, JAPAN

Miyashita, Yoshikatsu, Toyohashi University of Technology, Toyohashi, Aichi 440, JAPAN

Molino, Bettijoyce B., Office of Standard Reference Data, National Bureau of Standards, Rm B336, Bldg. 221, Washington, DC 20234

Moller, Eike, PH-E DB, Information u. Dokumentation /459, Aprather Weg, 5600 Wuppertal 1, WEST GERMANY

Moore, John W., Department of Chemistry, Eastern Michigan University, Ypsilanti, MI 48197

Moreau, Gilles, Roussel - UCLAF, 102, Route de Noisy, 93230 Romainville, FRANCE

Morrell, James A., Monsanto Co. - U2D, 800 N. Lindbergh Blvd., St. Louis, MO 63166

Mun, I., Lederle Laboratories, 500 N. Middletown Road, Pearl River, NY 10965

Neszmelyi, Andras, Hungarian Acad. Science, Central Res. Inst. Chemistry, 59/67 Pusztaszeri ut, H-1025, Budapest, HUNGARY

Neumann, David B., National Bureau of Standards, Bldg. 222, Rm B348, Washington, DC 20234

Newsome, Larry, US Environmental Protection Agency, OPTS/OTS/HERD/EEB (TS-796), Washington, DC 20460

O'Hara, Michael P., Questel, Inc., 1625 Eye Street NW, Suite 818, Washington, DC 20006

Opferkuch, Hans J., German Cancer Research Center, Im Neuenheimer Feld 280, 6900 Heildelberg, GERMANY

Patel, Siddhaarth M., The Goodyear Tire & Rubber Company, Goodyear Research, 142 Goodyear Blvd., Akron, OH 44316

Pauling, Peter, Chemistry Department, University College London, 20, Gordon Street, London,L, ENGLAND

Penca, Matej, Kemijski Institut Boris Kidric, Hajdrihova 19, YU-61000, Ljubljana, YUGOSLAVIA

Perkins, Miriam, Specialized Information Services, National Library of Medicine, 8600 Rockville Pike, Bethesda, MD 20209

Piottukh, Vadim, Chem., Org., Siberian Div. of Academy Sciences, 90 Novosibirsk 630090, USSR

Podgornaja, Rita, Inst. Org. Chem., Siberian Div. of Academy Sciences, 90 Novosibirsk 630090, USSR

Pool, M. Kay, Questel Inc., Suite #818, 1625 I Street NW, Washington, DC 20006

Potenzone, Rudolph Jr., US Environmental Protection Agency, MIDSD PM-218, 401 M Street, SW, Washington, DC 20460

Pretsch, Erno, Institute of Organic Chemistry, Swiss Federal Institute of, Technology, CH-8092, Zurich, SWITZERLAND

Qian, Cheng, Shanghai Institut of Organic, Chemistry, 345 Lingling Lu Road, Shanghai,, CHINA

Radak, Suzanne, Searle & Co., PO Box 5110, Chicago, IL 60680

Radermacher, L., Nuclear Research Center, Juelich GMBH, Dept. ZCH, P.O. Box 1913, D-5170 Juelich 1, WEST GERMANY

Randall, G.L.P., ICI Plant Protection Division, Jealotts Hill Research Station, Bracknell, Berkshire, ENGLAND

Rawson, Norman, IBM, 10401 Fernwood Road, Bethesda, MD 20817

Razinger, Marko, Boris Kidric Institute of Chemistry, POB 380, YU-61001, Ljubljana, YUGOSLAVIA

Rhodes, Yorke, New York University, Department of Chemistry, New York, NY 10003

Rosenthal, Donald, Department of Chemistry, Clarkson College of Technology, Potsdam, NY 13676

Rubenstein, Stewart, Harvard University, 12 Oxford Street, Box 100, Cambridge, MA 02138

Rzeszotarski, W. J., George Washington University, Department of Radiology, The University Hospital, 921 23rd Street, NW, Washington, DC 20037

Sasaki, Shin-ichi, Toyohashi University of Technology, Toyohashi, Aichi 440, JAPAN

Savage, Michael, Molecular Design, Ltd., 95 Madison Avenue, Morristown, NJ 07960

Schubert, Wolfgang, Org.-chem. Institut, Techn. Univ. Munchen, Lichtenbergstr. 4, 8046 Garching,, GERMANY

Shelley, Craig A., Eastman Kodak Company, Research Labs, B82, Rochester, NY 14650

Shue, Ho-Jane, Schering Corporation, B-8-2-29, 60 Orange Street, Bloomfield, NJ 07058

Shuely, Wendel, DRDAR-CLB-CP:Research Div., Chemical Systems
    Laboratory, Aberdeen Proving Grounds, MD 21010
Skidanow, Helene, Library L, Hoechst-Roussel Pharmaceuticals Inc.,
    Rte. 202/206 N, Somerville, NJ 08876
Smith, Stanley, Department of Chemistry, University of Illinois,
    Urbana, IL 61801
Song, Ban, Fein-Marquart Associates, 7215 York Road, Baltimore, MD
    21212
Spann, Mel, National Library of Medicine, Specialized Information
    Services, BISB, 8600 Rockville Pike, Bethesda, MD 20209
Sugnaux, Francois, Department of Pharmacology, Centre Medical
    Universitaire, 9 au de Champel, 1211 Geneve 4, SWITZERLAND
Suzuki, Isamu, The Tokyo Metropolitan Institute of Medical
    Science, Medical Engineering Division, 3-18-22 Honkomagome,
    Bunkyo-ku, Tokyo 113, JAPAN
Szekely, Gabor, Pharmazeutisches Institut, Universitat Bern,
    Ch-3012 Bern, Balzerstrasse 5, SWITZERLAND
Takayama, Chiyozo, Toyohashi University of Technology, Toyohashi,
    Aichi 440, JAPAN
Tobin, Frank, Fein-Marquart Associates, 7215 York Road, Baltimore,
    MD 21212
Uchino, Masahiro, Tokyo Institute of Technology, 4259
    Nagatsuda-cho, Midoriku, 227, Yokohama, JAPAN
Ugi, Ivar, Organisch-Chemisches Lab, Lichtenbergstrasse 4,
    Technische Univ. Munich, Garching 8046, WEST GERMANY
Ulyanov, G., Inst. Org. Chem, pr. Nauki 9, Academy of Sciences, 90
    Novosibirsk 630090, USSR
Varmuza, Kurt, Technical University of Vienna, Institute for
    General Chemistry 152, Lehargasse 4, A-1060, Vienna, AUSTRIA
Wang, C.D., Department of Chemistry, Univ. of North Carolina,
    Chapel Hill, NC 27514
Weber, Jacques, Department of Chemistry, University of Geneva, 30
    quai Ansermet, 1211 Geneva 4, SWITZERLAND
Wiggins, Gary, Chemical Information Center, Indiana University,
    Bloomington, IN 47405
Wilkins, Charles L., Department of Chemistry, University of
    California, Riverside, CA 92521
Wipke, Todd, Chemistry Department, Univ. of California, Santa
    Cruz, Santa Cruz, CA 95064
Wiswesser, William J., USDA, Weed Science Research, P.O. Box 1209,

Frederick, MD 21701

Wolman, Yecheskel, Organic Chemistry, Hebrew University of Jerusalem, Jerusalem , ISRAEL

Wood, William, US Environmental Protection Agency, OPTS/OTS/EED/CFB (TS-798), Washington, DC 20460

Woodruff, Hugh B., Merck Sharp & Dohme Research, Laboratories, P. O. Box 2000, Rahway, NJ 07065

Yamamoto, Takeo, University of Library & Info. Sci., Yatabe-cho, Tsukuba, Ibaragi 305, JAPAN

Yousif, Gariballa, Sir Elkhatim, Chemistry Department, Faculty of Science, University of Khartoum, P. O. Box 321, SUDAN

Ziegler, E., Max-Planck-Institut, Kaiser-Wilhelm-Platz 1, D 4330 , Mulheim/Ruhr, WEST GERMANY

Zoebisch, Eve, University of Texas, Chemistry Department, Welch Hall 5.336, Austin, TX 78712

Zupan, Jure, Kemijski Institut Boris Kidric, Hajdrihova 19, YU-61000, Ljubljana, YUGOSLAVIA

ELECTROPHORESIS

A Survey of Techniques and Applications

edited by Z. DEYL, Czechoslovak Academy of Sciences, Prague

JOURNAL OF CHROMATOGRAPHY LIBRARY, 18

PART A: TECHNIQUES

Z. DEYL *(editor)*
F.M. EVERAERTS, Z. PRUSÍK *and*
P.J. SVENDSEN *(co-editors)*

"… provides a sound, state-of-the-art survey of its subject".
– Chemistry in Britain

"… the editors have set out to bring everything together into a coherent whole… they have succeeded remarkably well… the book is bound to be well liked and appreciated by readers".
– Journal of Chromatography

This first part deals with the principles, theory and instrumentation of modern electromigration methods. Both standard procedures and newer developments are discussed and hints are included to help the reader overcome difficulties frequently arising from the lack of suitable equipment. Adequate theoretical background of the individual techniques is given and a theoretical approach to the deteriorative processes is presented to facilitate further development of a particular technique and its application to a special problem. In each chapter practical realisations of different techniques are described and examples are presented to demonstrate the limits of each method.

CONTENTS:
Introduction. Chapters: 1. Theory of electromigration processes *(J. Vacík)*. 2. Classification of electromigration methods *(J. Vacík)*. 3. Evaluation of the results of electrophoretic separations *(J. Vacík)*. 4. Molecular size and shape in electrophoresis *(Z. Deyl)*. 5. Zone electrophoresis (except gel-type techniques and immunoelectrophoresis) *(W. Ostrowski)*. 6. Gel-type techniques *(Z. Hrkal)*. 7. Quantitative immunoelectrophoresis *(P.J. Svendsen)*. 8. Moving boundary electrophoresis in narrow-bore tubes *(F.M. Everaerts and J.L. Beckers)*. 9. Isoelectric focusing *(N. Catsimpoolas)*. 10. Analytical isotachophoresis *(J. Vacík and F.M. Everaerts)*. 11. Continuous flow-through electrophoresis *(Z. Prusík)*. 12. Continuous flow deviation electrophoresis *(A. Kolin)*. 13. Preparative electrophoresis in gel media *(Z. Hrkal)*. 14. Preparative electrophoresis in columns *(P.J. Svendsen)*. 15. Preparative isoelectric focusing *(P. Blanický)*. 16. Preparative isotachophoresis *(P.J. Svendsen)*. 17. Preparative isotachophoresis on the micro scale *(L. Arlinger)*. List of frequently occurring symbols. Subject Index.

1979 xvi + 390 pp.
ISBN 0-444-41721-4

PART B: APPLICATIONS

Z. DEYL *(editor)*
A. CHRAMBACH, F.M. EVERAERTS *and*
Z. PRUSÍK *(co-editors)*

Part B is an exhaustive survey of the present status of the application of electrophoretic techniques to many diverse compounds. Those categories of compounds most suited to these separations, such as proteins and peptides, are dealt with in detail, while the perspectives of the applications of these techniques to other categories of compounds less commonly electrophoresed are given. Special attention is paid to naturally occurring mixtures of compounds and their treatment. This is the first attempt to cover the field on such a broad scale and the book will be valuable to separation chemists, pharmacologists, organic chemists and those involved in biomedical research.

CONTENTS: 1. Alcohols and phenolic compounds *(Z. Deyl)*. 2. Aldehydes and ketones *(Z. Deyl)*. 3. Carbohydrates *(Z. Deyl)*. 4. Carboxylic acids *(F.M. Everaerts)*. 5. Steroids and steroid conjugates *(Z. Deyl)*. 6. Amines *(Z. Deyl)*. 7. Amino acids and their derivatives *(Z. Deyl)*. 8. Peptides and structural analysis of proteins *(Z. Prusík)*. 9. Gel electrophoresis and electrofocusing of proteins *(edited by A. Chrambach)*. Usefulness of second-generation gel electrophoretic tools in protein fractionation *(A. Chrambach)*. Membrane proteins, native *(L.M. Hjelmeland)*. Membrane proteins, denatured *(H. Baumann, D. Doyle)*. Protein membrane receptors *(U. Lang)*. Steroid receptors *(S. Ben-Or)*. Cell surface antigens *(R.A. Reisfeld, M.A. Pellegrino)*. Lysosomal glycosidases and sulphatases *(A.L. Fluharty)*. Heamocyanins *(M. Brenowitz et al.)*. Human haemoglobins *(A.B. Schneider, A.N. Schechter)*. Isoelectric focusing of immunoglobulins *(M.H. Freedman)*. Contractile and cytoskeletal proteins *(P. Rubenstein)*. Proteins of connective tissue *(Z. Deyl, M. Horáková)*. Microtubular proteins *(K.F. Sullivan, L. Wilson)*. Protein hormones *(A.D. Rogol)*. Electrophoresis of plasma proteins: a contemporary clinical approach *(M. Engliš)*. Allergens *(H. Baer, M.C. Anderson)*. 10. Glycoproteins and glycopeptides (affinity electrophoresis) *(T.C. Bøg-Hansen, J. Hau,)*. 11. Lipoproteins *(H. Peeters)*. 12. Lipopolysaccharides *(P.F. Coleman, O. Gabriel)*. 13. Electrophoretic examination of enzymes *(W. Ostrowski)*. 14. Nucleotides, nucleosides, nitrogenous constituents of nucleic acids *(S. Zadražil)*. 15. Nucleic acids *(S. Zadražil)*. 16. Alkaloids *(Z. Deyl)*. 17. Vitamins *(Z. Deyl)*. 18. Antibiotics *(V. Betina)*. 19. Dyes and pigments *(Z. Deyl)*. 20. Inorganic compounds *(F.M. Everaerts, Th. P.E.M. Verheggen)*. Contents of "Electrophoresis, Part A: Techniques". Subject Index. Index of compounds separated.

1982 xiii + 462 pp.
ISBN 0-444-42114-9

ELSEVIER
P.O. Box 211, Amsterdam
The Netherlands

52 Vanderbilt Avenue
New York, NY 10017, U.S.A.

7255